SEA LEGS

SEA LEGS

Tales of a Woman Oceanographer

KATHLEEN CRANE

Westview
PRESS

A Member of the Perseus Books Group

Westview Press books are available at special discounts for bulk purchases in the United States by corporations, institutions, and other organizations. For more information, please contact the Special Markets Department at the Perseus Books Group, 11 Cambridge Center, Cambridge MA 02142, or call (617) 252-5298 or (800) 255-1514 or email j.mccrary@perseusbooks.com.

Published in 2003 in the United States of America by Westview Press, 5500 Central Avenue, Boulder, Colorado 80301–2877, and in the United Kingdom by Westview Press, 12 Hid's Copse Road, Cumnor Hill, Oxford OX2 9JJ.

Find us on the World Wide Web at www.westviewpress.com

A Cataloging-in-Publication data record for this book is available from the Library of Congress.

ISBN 0-8133-4004-7 (HC)
The paper used in this publication meets the requirements of the American National Standard for Permanence of Paper for Printed Library Materials Z39.48–1984.

Interior book design by Lisa Kreinbrink
Text set in 12-point Dante MT

10   9   8   7   6   5   4   3   2

FOR DAVID

There are four winds and eight subwinds
Each with its own colour
The wind from the East is a deep purple,
From the South a fine shining silver.
The North wind is hard black and
The West is amber.

—FLANN O'BRIEN

CONTENTS

ACKNOWLEDGMENTS

Twenty-nine years have passed since I first began my explorations in oceanography. As I reflect over these decades, I realize that my struggle to surmount the difficulties that other women and I faced in science during the Cold War have not been for naught. Today many more women work in oceanography than ever before. Although some of my colleagues played a role in maintaining the Cold War, many others worked towards its' dismantling. It is my hope that the collective efforts of women and men who have encouraged communication, amongst themselves and between nations, will lead to a more humane and equitable society, in which both science and the spirit are equally treasured.

I am indebted to Scripps Institution of Oceanography, Woods Hole Oceanographic Institution, Lamont-Doherty Earth Observatory, and Hunter College for giving me the opportunity to explore the oceans, discuss issues, and develop and write the chapters of this book. Thanks go to the many colorful individuals who laid the foundations of the early years of my life in oceanography. Some of these people are mentioned in this book; unfortunately there was not enough space to mention many others. Interactions with my students at Hunter College stimulated many of the thoughts presented within, ranging from the role of the military and politics on the evolution of science, to politics' and society's role on the changing geography of gender.

Certain individuals have provided me with insight to the workings of the Former Soviet Union and of the United States during the Cold War period. These individuals, who do not wish to be named, are everywhere in the book in spirit.

I thank my editor Holly Hodder for helping me to craft these ideas into chapters and for convincing me to write this story. She and Barbara Greer at Westview Press were very indulgent of my whims and corrections.

I give a special thanks to Frank Aikman, Caroline Baum, Enrico Bonatti, Georgy Cherkashov, Mike Coffin, Godfrey Day, Joan Gardner, Svetlana Gataulina, Else Haugland, Rachel Haymon, Tom Jordan, Leslie Karsner, Heidi Kassens, Bob Kieckhefer, Bob Kleinberg, Emory Kristof, Cindy Lee, Ken Macdonald, Jurgen Mienert, Suzanne O'Connell, Helmut Plant, Kathy Poole, Vincent Renard, Alexander Shor, Anders Solheim, Ralph White, and Karen Wishner who shared many kinds of adventures with me, through thick and thin, all over the world.

I thank my family, and my daughter's godparents and sitters, including Mena Behsudi, Ted Dengler, Leslie Paine, Nancy Steelberg, and Nancy Templeton for their double duty in managing my household and tending my daughter during my many weeks of absence. Without them I could not live my life as both a parent and as a scientist.

Finally, I thank my daughter, who shows me every day what is important in this life.

Kathleen Crane
McLean, Virginia
November, 2002

INTRODUCTION

~

Aghast at what had been done with the basic equations
on the conversion of mass into energy, physicist Albert
Einstein shortly before his death went on the air to call
the U.S. and the U.S.S.R. "hysterical." Their arms race, he
said, "beckons annihilation. The first problem is to do
away with fear and mistrust."

—JOHN STUART MARTIN

It was the winter of the year 2000. My daughter was playing with
her paints on the living room table. Red and purple splashed over
the paper when the phone rang. The rooms in our house were in a
state of chaos, with scarves, mittens, heavy boots, books, and papers,
as we were readying to make a long journey into the Russian winter.
Our apartment and nanny were waiting for us in St. Petersburg. My
Russian and German colleagues were also waiting. Our Russian auto-
mobile, a Lada, was waiting, the World Wildlife Fund was waiting,
and maybe even the Russian Federal Security Service (the FSB, formerly the KGB) was waiting. All of this was waiting, and the pressure
was growing heavier. I went to the phone, worried with premonition.

"Dr. Crane, this is the FBI." My heart sank and I could hardly
answer.

"Dr. Crane, we want to speak with you in person, about your
acquaintances, about your Russian acquaintances." My heart sank
even more. It was too much.

In October, Josh Handler, who worked on environmental issues for Princeton University, had been questioned in Moscow for ten hours by the FSB. His Russian colleague had been arrested on charges of espionage, and Russian officials had confiscated their computers. Many messages from me populated files on Josh's computer. If I went to Russia, would I also be harassed? Would I be a sitting duck because I had the right connections at the wrong time? Had the wrong people funded my work?

For seven years I had worked with a group who tried to bring oceanographers from the former Soviet Union closer to oceanographers from the United States, first by way of funding allocated by the U.S. Congress through the Office of Naval Research and later through the World Wildlife Fund. Our goals were twofold: to develop environmental and oceanographic programs and to create an atlas of contamination in the Arctic Ocean and its surrounding areas. Scores of scientists from the former Soviet Union had been hired; many of them had visited the United States, and some had stayed in my home.

Piles of purely oceanographic data had been rescued from the dark halls of Soviet institutes slated for imminent closure, and even research ships had been funded. Finally, Americans were able to work together with Russians on the high seas along the mid-ocean ridge on the bottom of the Atlantic Ocean and throughout the Arctic Ocean. It was good for Russia, it was good for America, and it was good for the world.

What had gone wrong seven years later? Why had our programs, lauded in both countries, become suddenly suspect? The talks between Vice President Al Gore and Russian Prime Minister Viktor Chernomyrdin had reached a stalemate. The unending problems in the former Yugoslavia were baring their nasty teeth. The Russians were allowing centuries of xenophobic paranoia to rekindle itself, and the U.S. government sought to find a demon of espionage in every government laboratory. Those of us who had worked to bring Russia and America closer together were suddenly being scrutinized. It seemed as if we were walking on a very long diving board that suddenly might be sawed off behind us, leaving us to tumble into the rising tide of distrust.

Why were people and their governments so suspicious about what we were doing? We worked as scientists, attempting to make the world a little bit cleaner than the one into which we were born.

In the early 1990s, much of the scientific and environmental information collected by the former Soviet Union was declassified and made available to researchers worldwide. For the first time in decades, Russians could access their own data. But information is often viewed as dangerous, and in the case of Russia—a country responsible for a disproportionate amount of Arctic Ocean contamination—the newly declassified environmental research painted a negative picture. Much of that information was now locked away, designated as classified once again.

Just a few months earlier, in the fall of 1999, Alexander Nikitin, an outspoken Russian environmentalist, had been tried for treason. Although he was acquitted, the Russian government was appealing the decision. The Russian press was claiming that environmental organizations were filled with foreign spies.

I was getting very nervous. I had secured a contract with the World Wildlife Fund to go to Russia during my sabbatical leave to update and produce an environmental atlas of the Arctic region in Russian for distribution in Russia. What did this rising tide of unrest mean for me and my colleagues in the United States and in Russia?

Were we suddenly considered to be international spies? What would happen to the study of oceanography in Russia if funding from the outside world ceased? What would happen to my colleagues? Might they be considered guilty by association with me?

What about my daughter? Would something terrible happen to her if I took her back to Russia, the land of her birth?

In October 1999, after I had discussed the situation with Josh, I delayed my journey. In November, I delayed it again. December came, followed by the millennium, and I found myself still in Washington and still afraid.

The FBI agent on the phone jostled me back to the present. "Dr. Crane, we understand you have many acquaintances from Russia. When may we speak? Don't you want to help your country?" I wished instantly that I was back on the bottom of the ocean, where no one could interrogate me, a region far removed from the prying questions of any government.

This distrust was not new to my profession, however. Ironically, the science of oceanography had been driven by military competition. I was born in 1951, a child of the Cold War, and I, like others of my generation, was encouraged to pursue a career in science, largely in response to the politics and fear that emanated from that nearly fifty-year-long era.

I have written this book to bear witness to the effects of the Cold War on the science of oceanography and those who have pursued its study. During the second half of the twentieth century, complicated human and political interactions created and destroyed the careers and personal lives of women and men. The intense changes that our society experienced allowed fantastic discoveries in the sea, yet the raw passion that erupted during this period unfortunately buried many others.

I went to sea during the start of the profound changes that shook the world of oceanography. I was a firsthand witness in the deep-sea trenches and a captive of the Cold War. The story is one of beauty and pain and of liberation and change.

1

DUCK AND COVER

The most portentous event of 1954 was the death of
a humble Japanese fisherman. His name was Aikichi
Kuboyama and he was the first victim of a hydrogen
bomb. Kuboyama was fishing in the Pacific Ocean when
he heard "the sound of many thunders" and saw "flashes
of fire as bright as the sun itself." What he heard and saw
happened 80 miles away. . . . For every adult in the world,
the death of this simple man dramatized the H-bomb's
terrifying power and unspeakable possibilities of a hydro-
gen-era war. . . . So staggering a prospect set the best
brains of the free world struggling to avert it. Throughout the year the statesmen met in London, Paris, Brus-
sels, Geneva, Berlin. But what was being thought behind
the Iron Curtain, no one could say for sure.

—RICHARD M. GORDON, EXECUTIVE EDITOR,
THE UNICORN BOOK OF 1954

It is a cold spring in Arizona. Dust blows, sculpting the parched
landscape. In one glimpse, I can view the volcanic peak of San
Francisco and, in the foreground, the pungent sage bending under
the desiccating winds. The wide blue sky hovers above. Then my
memory peels back to other skies, unfathomable distances, where
little evidence of life can be seen. Blue rolls into fog, into brilliant
star-studded nights, and into the pink glow of a delicate sunrise.
Below, there is no sagebrush, no dust. There is no evidence of land
at all. It is the most common view on Earth. It is the sea. These

visions haunt my days and nights. I cannot control when they will tumble back to me, awake or asleep, on a hike in the desert or at my desk in Manhattan.

My oceanographic career spans six oceans, from the Arctic to the tropics, from the sea surface to the depths of the mid-ocean ridge and beyond. It encompasses more than four years at sea, on vessels operated from many nations—ships, pontoons, and submarines—through days and weeks of little sleep, through storms and tropical calm.

Modern oceanography emerged from the battlegrounds of World War II, when the world's navies realized how little they knew about the sea and its depths. Knowledge of the sea was crucial for everything from submarine warfare to landing troops on the coral-rimmed beaches of the Pacific islands. Then in the 1950s a new war emerged. It was an insidious conflict of hostile rivalry, filled with stealth missions in the deep-sea abyss and dominated by the threat of nuclear weapons. Oceanographers were hired to monitor atomic blasts at South Pacific islands and atolls. The developing knowledge of the sea, its life, and the seafloor below was used to determine "battle space" environments in the ocean.

My future was defined on October 4, 1957, when the Soviet Union successfully launched the *Sputnik I* satellite into space. I was six years old and had just started the first grade. Although the *Sputnik* launch was a single event, it triggered the start of the space age and the U.S.-U.S.S.R. race for the conquest of space. *Sputnik* had caught the American public off-guard. The United States feared that the Soviets' ability to launch satellites translated into the ability to launch ballistic missiles that could carry nuclear weapons from Europe to America. Senator Lyndon Johnson, alarmed at the Russians' big lead in the space race, exclaimed, "Soon they will be dropping bombs on us from space like kids dropping rocks onto cars from freeway overpasses." Newspapers ran bold headlines: "Reds Orbit Artificial Moon."

Then the Soviets struck again. On November 3, 1957, *Sputnik II* was launched, carrying a much heavier payload, including Laika, a live dog.

Engineering colleges were flooded with new students the following year. It was as if everyone were "joining the army" to take on the Russians in the new frontier of space. Politicians and pundits attacked the U.S. educational system for falling behind the Soviet schools in training students in the sciences. Suddenly, in response to the demand to swell the ranks of the scientific community, the doors were opened to a few select women to study science and mathematics during the late 1950s and 1960s. I would benefit from *Sputnik* and the consequent fear it sparked in America; because of the critical shortage of men in scientific fields, I was encouraged to become a scientist. But the inclusion of women in science during the 1960s and 1970s was always an "on again, off again" sort of story. Were it not for the threat of the Soviet Union, I am sure that I would not be a scientist today, because science had been dominated by men through much of its history.

Both the Americans and the Soviets built massive armies, navies, and air forces and prepared their scientists for a global nuclear war. I wanted nothing to do with global war, and I certainly did not want to become another physicist guiding missiles. Instead, I had fantasies of adventures and explorations to parts unknown, uncivilized, and uncharted. I wanted to see the whole world and be a part of it instead of engaging in fear and promoting national isolation. But I kept these feelings to myself. It was obvious that I would only be scorned if I voiced my opinions. This was something I had learned from my very private family.

My parents were from the Midwest, and they embodied the ideals of strength and privacy. Our family never had a lot of money; every penny we earned was put straight into the bank for college. For our allowance we received twenty-five cents every two weeks, and we were supposed to save it, not spend it. Friends were not encouraged to visit (there were already eight of us in the house), and so our exposure to the outside world was limited. But our home was filled with books; books were our virtual pathways out of suburbia. As a family, we traveled only to Michigan, where my grandfather owned a small cottage at a lake. Otherwise, my family managed to invade the outside world with our walks and picnics. We were truly a unit unto ourselves.

My parents were, I guess, an odd couple, yet theirs is a love story hard to match. My father, a singer, left his radio show in Chicago in the late 1940s to marry and move east with my scientist mother. In fact, they were fleeing two very disapproving families. At the time of their marriage, each of my parents already had one child in tow, and so their union began as a family of four. During World War II, my mother had worked as a dietician at the University of Chicago, while her first husband was at war in the Pacific. Their marriage faltered after his return, probably as a result of the changing role of women in society. During this same time, my father had suffered enormous tragedy with the loss of his wife, sister, and father all in the same year. Yet, somehow these grief-stricken war years ended in unexpected happiness. My parents married, started new lives in Washington, D.C., and created a new family of four additional children. Although my mother willingly gave up her professional career when she married my father, I always wondered if she felt a sense of loss during the overwhelming responsibilities of motherhood. Curiously, even though she was a very independent woman, my mother pushed her sons much harder than her daughters to become scientists, doctors, or engineers.

When Bikini Atoll was blown to smithereens by nuclear testing, witnessed by legions of sailors and oceanographers in the Pacific, I was a little girl, hiding my head in the "duck and cover" position under my school desk in the suburbs of Washington. Every week we were shown movies of nuclear war and instructed in how to save our lives if the Soviets attacked. In case of atomic war, my father, who worked for the U.S. Post Office, was to evacuate to Winchester, Virginia, where the U.S. government would set up emergency operations, far from the Capitol. I remember him saying, "If an attack happens while you are at school, you must find your way to Winchester. That is where I will be." I was seven years old and terrified. My childhood was so colored by the fear of nuclear attack that I wanted only to escape.

The door that *Sputnik* opened for young girls didn't remain open for very long, and many of the girls in my class were caught between the conflicting expectations of government and society. Some would become victims of a world oscillating between the Cold War fears that created new opportunities for women and the traditional gender-segregated society that characterized suburban America in the 1940s and 1950s.

I remember thinking when I was only six years old that our society was really unfair to girls. My brother David and I were close friends and only a year apart, and I often felt that he needed me by his side, yet he was allowed to ride his bicycle to the next neighborhood and I was not. I was just as strong—perhaps stronger. I certainly was better at baseball, and I was quite talented with hammers, nails, chisels, and saws. I made my own bows and arrows and built stick houses. I was even a "leader" on the street because I could jump higher and run faster than the others in the pack of neighborhood children. When I reached the age of seven or eight, I began to dream of being an explorer like Daniel Boone, Tarzan, Robin Hood, and the Swamp Fox. To me, they were symbols of the successful struggle for freedom. I loved reading about their exploits, yet the "real" world told me that as a girl I would never be able to enjoy their adventures. Books spoke of exploration, while society spoke of becoming a housewife. The pressure from society paralyzed me, and by age eleven, I felt as though I were fighting for my life.

While my school was subtly channeling me into science, I grew further and further away from most of my classmates, and so I was always alone. In my solitude, I built my dreams and fantasies of escape. During the evening hours, I used to lie in front of a series of old painted maps, which my mother had purchased from the Golden Gate International Exposition of 1939 and 1940. I remember the Pacific Ocean as a glorious blue expanse, dotted with small exotically painted tropical boats and native peoples. As the years went by, I swam into these evocative maps. I became a Polynesian; I herded sheep in Australia and spearfished in Tahiti. These maps, framed by fragrant bamboo, were my visual escape, a door to a geographical freedom. I dreamed of exploring new islands, mapping new places, discovering new peoples and their customs, at home in new cultures, and swimming every day in an azure sea. Why had my mother bought these maps? Perhaps she, too, fantasized about journeys to faraway places, unimpeded by mid-twentieth-century American life. To this day, I don't know if this is how she felt, but I imagine it to be true.

My life as an oceanographer began in the fantasies I had built around my mother's maps. However, my search for adventure and my attempt to escape from the fears of the Cold War led me straight to those I thought I was retreating from.

2

VICTORY AT SEA

Moon River, wider than a mile,
I'm crossing you in style some day.
—HENRY MANCINI

When the blue winds blow gently over the slate sea, palms
rustle, the soft warm raindrops tumble down, the world of
the mind drifts away, and the body melts into the elements. Memories, supple and bright, evoke times on ships, gently rocking, gently swaying under a tropical heat. Stars revolve in a velvet black sky,
while the warmth of a caressing hand lengthens the moments
stolen from hard labor. These reveries are always there in the background of the everyday world.

In the tropics, the blue sky squeezes the rays of the sun through
the wafting palm trees. Their shadows, thin gray slivers, edge
across amber-tiled floors. The warmth, the blue drift in and out like
wave tops on an infinite, deep sea. This is the open world, free in
motion, subtle and mysterious, but scratch its surface and there are
stories of incredible vastness, imbued in life and transformation.
There is no death, no disappearance, only a transmutation from
magma and vapor to rock and sea, soil and sky, and from the sky to
the universe beyond.

The romance of the changing Earth, planted in my imagination
through science and fiction, lured me to wild and unmapped

places. I used to watch the stories of *Victory at Sea,* rife with wild soaring music and waves breaking over aircraft carriers that were steaming deep into the South Pacific. These battle stories from Midway, Truk, New Guinea, and Japan haunted me, their ticking Morse code echoing in my ears. I could imagine myself there on the flight deck directing planes, navigating through rough waters. Communication in staccato—acoustic undulations, frequency modulations, the sounds of naval sonar—reverberated in my head. Two ships down, torpedoes away, contact, explosion.

My eventual foray into decoding the deep was only one step from such romantic visions. This fascination with conquering and deciphering something unknown would eventually develop into real goals. It was fun, and it was a challenge: to attack something and win, to make a trace along the seafloor, to track sound through the sea, to whip a ship around on its cycloids, or to snake a bottom-towed instrument in and out of a deep-sea volcanic crater. Of course, the hell of war portrayed in *Victory at Sea* never occurred in my own life. Instead, I would know the comradeship of ships' crews and victory over difficult obstacles. I would come to understand the real-life romance of living through challenging times.

Throughout high school, I kept these dreams to myself. I studied math and science and spent most of my time reading, shut away from my family. My dreams eventually led me to the West Coast of the United States, to Oregon. The year was 1969. Although the United States was officially fighting the war in Vietnam, young people across the country were fighting a different war, asserting their rights, seeking liberation from "the establishment." I, too, was seeking liberation. I had resolved to become an explorer, and as a woman, this meant I would spend years in college earning advanced degrees.

My goal was to major in a subject that would lead me to oceanography, which I could not pursue until graduate school. I chose geology as my undergraduate major because I wanted to spend as much time as possible in the outdoors. At the time, U.S. geologists were mostly very conservative "back-country" men. They fostered anti-intellectual attitudes, were heavy drinkers, and grew big, bushy beards (or at least this was my impression). The geologists at Oregon State University were no different. I entered as the first female undergraduate in the department. The resentment that this provoked was almost unimaginable. But anger was everywhere then, mostly as an expression of fear about the changes rippling throughout our society. It permeated every facet of my college life.

As a geologist, I hoped to examine the big picture of the Earth and to link its physics to its beauty. At that time, in that department, that would prove to be impossible. At least I thought it was impossible, until I discovered the theory of plate tectonics, purely by chance. One day, I was flipping through a reference book for a paleontology paper I was writing, when I spotted some graphics that depicted the bottom of the ocean, its mid-ocean ridges, and its moving and subducting plates. I couldn't stop reading. The static Earth of my textbooks seemed moribund in comparison. The theory of plate tectonics portrayed a living planet, always moving, always seething. Were continental collisions responsible for mountains? Were fault valleys created in response to the gradual spreading apart of landmasses—continental rifting? The more I read, the more I had to read. This was the first exciting story I had come across in geology, and the experience of discovery felt like I was learning to fly. The chapters in that reference book held my attention for a week. Out of my discoveries grew a paper that attempted to tie together a synchronous pattern of mountains building across the world as a result of global crashing and bashing. I handed it in with great confidence. Two weeks later, the paper was returned, but not with an "excellent" scrawled on the top. Instead, above my name the words glared, "This Theory of Plate Tectonics is total bunk." How could this be? I had spent days in the library devouring books and searching for clues to the mystery of my topic. I received a grade of D.

This was February 1970. To recover, I fell back into the secure and nonthreatening routine of identifying rocks and peering down microscopes. When possible, I stole away to be with my colleagues in German studies or my friends in the physics department. Gradually, the excellent grades accumulated: A's in physics, calculus, literature, and history. Only the onerous D in geology remained, and it was my major. Shortly after the disconcerting D episode, I was assigned a new advisor, Julius Dasch, newly arrived from Yale University. He was young, bright, energetic, and clever. He pushed me on, defended me, and gave me hope. Dr. Dasch believed in plate tectonics, which left him quite alone among the staff. He was a liberal upstart who had more connections with the School of Oceanography than with the Department of Geology. It was through him that I found my way into the field of marine geology. I became entranced with the fact that although the seafloor comprises more than 70 percent of the Earth's surface, most geology departments refused to have anything to do with this vast domain. They willingly wore blinders when it came to the study of the Earth.

My next frustration with the geology department arose when the faculty chose a student for the yearly Union Oil Company Scholarship. Dr. Dasch pushed for me to receive this honor. I had the highest average among the eligible students. When Dr. Dasch called me into his office, explained to me that many of the professors in the department had written a letter to Union Oil in California requesting that the scholarship not be awarded to a woman, I was dumbfounded, shocked. I was wholly unaware that the sentiment against women still prevailed. However, Union Oil did not accede to their wishes. The scholarship was awarded, and I, the first woman in the department, became its dubious recipient. I grew tired of these people and this department. It was time to get out.

3

WORLD GEOGRAPHY

Auf der Fortuna ihrem Schiff
  Your ship called Fortune
Ist Er zu segeln im Begriff
  Is ready to sail;
Die Weltkugel liegt vor Ihm offen
  The globe lies wide open before you,
Wer nichts waget, der darf nichts hoffen
  Who never dares, should never dare to hope.
—Friedrich von Schiller, *Wallenstein*, 1798

I recently visited Carol Ruekedeschel, a biologist who lives on the remote northern end of Cumberland Island, off the southern coast of Georgia. Gatherer of dead armadillos, mice, rats, otters, mink, birds, alligator scat, and sea turtles, Carol had lived here for close to twelve years, building and rebuilding various structures and dwellings. First she built a house of weathered wood, then a storage shed, then a "nerve center"—a small laboratory filled with formaldehyde-drenched specimens, shells, and indescribably foul smells of decay and preservation. Scattered around the structures lay piles of boards, rusted buckets, iron kettles, patches of bolting lettuce plants, frost-tinged daffodils, and an algae pool into which a garden hose trickled. Cats, hens, roosters, and turkey vultures roamed the terrain. Carol Ruekedeschel, naturalist, scion of Cumberland Island, is here because she wants to be.

In 1970, my goal was to be like Carol. I was tired of academics, fed up with people, and I dreamed of retreating into the forest. I wanted to be a fire ranger or a naturalist, out of the office and into the trees. My goal was to find a field of action with the right combination of mind and body. I would go anywhere and do anything.

The year wore on, class after class. Then I found a discipline I felt at home in, and it was physics. This was a bit unexpected, yet I found that the professors in this field were apparently blind to gender and seemed delighted that I enjoyed writing equations. They were lost in their own ionosphere of sorts, and I didn't worry about feeling different or somehow threatening. In any case, I realized that physicists could be separated into two camps: those who were essential to the Cold War arms race, and those who were humanists vehemently opposed to the war. In fact, it was an outspoken group of physicists who kept the lines of communication open between the Soviet Union and the United States throughout the Cold War. I certainly was not driven by the Cold War, nor did I wish to work in a top secret, closed-door environment.

One of my professors, Dr. Olaf Boedtker from Switzerland, was adept at rapidly covering the blackboard with equations, yet he still managed to inspire his students to follow fields unrelated to physics. He knew that I was interested in international events, and he persuaded me to apply to the university overseas study program in Germany. I was reluctant, but I applied and was accepted.

That summer I taught swimming at Lake Michigan near my family's cottage. I did not look forward to the upcoming year in Germany, yet I wanted to live there for several complex reasons. I was intrigued by human nature, and I wondered how the Germans endured after the dark, vilifying years of World War II. I also wanted to confront the Berlin Wall that physically and emotionally divided East from West.

In the beginning, my German escapade was a journey into fear and deep loneliness. I worked so hard to fit in that unless I could express myself in German, I never uttered a word. Two or three months passed like this, but then I gradually did learn to speak German. I lived with Germans, and I felt German. To satisfy my education requirements, I enrolled in geology, German literature, and the history of Western civilization at the University of Stuttgart. The geology we studied focused on the European Alps, and I learned millions of trivial geological facts that could not be tied together because the unifying theory of plate tectonics had not yet been accepted in Europe outside of England. However, my real learning experience came not from the science I studied but rather from the international politics I experienced. In class we examined the historical expansion of the Arab world and the role of the Monroe Doctrine from the German point of view, and this outlook became a guidepost in my later life.

As I came to feel German, I felt ashamed. I felt the animosity from the French, the Belgians, and the Dutch whenever I traveled with fellow Germans into those countries. I had stepped into someone else's skin, and I could feel the pain of isolation. Was this a precursor, a training ground for the isolation of my future role as one of the first women in oceanography? Did it sow the seeds for my later work with the Russians? The discomfort I felt in traveling abroad with Germans taught me compassion for the vanquished.

Europe in 1971 was a battleground between East and West. Rebuilding society after the devastation of World War II was still ongoing in the West, yet across the Berlin Wall lay a society frozen in the grip of time. At this site, the tyranny of Soviet Stalinism stood in stark opposition to the rampant capitalism of the West. The memory of unspeakable human violence lay like a shroud over the populations of both East and West, and it began to affect me as well.

After a time, I yearned for the peace of the natural world. I ached for the ocean, for its swash and backwash, miles of pure soft sand, and undulating dunes spiked by russet grasses. I wanted to feel a beach, not a German city. I wanted to feel the hot morning sun as it pierced my sweater, stretching my shadow long across the sand. I wanted the gulls' plaintive cries over the crashing surf, and I wanted the sunrise to be milky, slowly diffused out of a rising band of light.

By January, I needed to escape from Germany. I ambled over the Alps, visiting much of Europe and the Mediterranean, seeking the open and free sea. No place within the sphere of Europe was too distant or too difficult to reach. In 1971, the geography of the world lay at my feet: Italy, Greece, and then Turkey. My plane ticket was purchased through the sale of one quart of my rich red blood. I arrived in Izmir, rode by cart, in old Chevrolets, and in slow trains until days later I traveled by boat across the Sea of Marmara to Istanbul. Once there, Haluk, a Turkish friend, led John, an architect friend, and me into the depths of political despair. Haluk turned out to be a student activist, a harmless occupation normally, but not in Turkey in 1971. Again and again, we were accosted by government-supported gangs outside of restaurants. Nightly, thick-armed thugs leapt out of the street's black shadows, throwing John and me aside, leaping onto Haluk and his friends. Fist after fist pummeled into these people until they hit the pavement. Then, as quickly as they appeared, the thugs would vanish.

In 1971, the student activists, however peaceful, were seen as a threat to the tenuous Turkish government. The government intended to provoke them into fighting back, and once a student lifted a hand, the thugs would shoot to kill. That year, the military hanged five Turkish students. All universities were closed, the best and the brightest imprisoned. Those who questioned authority disappeared. These tragic circumstances killed any vestiges of political naïveté I may have retained. At the age of nineteen, I found myself in the middle of tyranny.

How strange it was then to leave Turkey and to find myself in the urbane seaside metropolis of Monte Carlo. The place repelled me with its shining high-rises but attracted me, too, because ensconced within the prince's enclave was the thriving royal oceanographic museum and research center (Le Musée Océanographique de Monaco). This research center was the first of its kind in the world. Its saltwater aquarium housed many species of fish from the Mediterranean Sea, and in the middle of the museum was Jacques Cousteau's yellow diving saucer, which hung gracefully in the central exhibit hall. This fascinated me most of all. I had yearned to see this submarine since I had first encountered it through one of Cousteau's movies, which he presented to the National Geographic Society in Washington, D.C. (Again, it was my mother who had taken me to Cousteau's lecture; I don't think I was older than ten.) Since that encounter, I had been longing to meet Cousteau himself. When I asked a member of the oceanographic museum staff whether Cousteau was in residence, I was told that Cousteau spent most of his time in the United States, because Americans donated more money to his projects than the French. It probably didn't really matter, because the bug of oceanography had already bit me hard. When I arrived back in Stuttgart, I let my director, Helmut Plant, know that perhaps I would leave the overseas studies program to seek a job in Monte Carlo at the oceanographic museum. I was anxious to begin the next chapter in my life. However, I didn't go to Monte Carlo; I stayed and completed my year in Germany.

In May of that year, I traveled to Eastern Europe. The Iron Curtain defined all of our lives in Germany. Bullet holes in my university buildings gave daily testimony to the reality of both World War II and the Cold War in Europe. Massive new construction in the West was juxtaposed against the gray and silent East. I traveled to East Berlin with a group of foreign students, and I felt firsthand the utter darkness of isolation. I felt sick, passing through Checkpoint Charlie, the political boundary between West and East, but I was also drawn to the plight of those enmeshed in this devastating situation. I carried packages to families in East Berlin from their West German friends and families: stockings, oranges, and bananas. In East Berlin, I saw for myself the goose-step march of the soldiers stationed there. I was shocked to learn that my West German friends had little or no knowledge that this march was a remnant of the Nazi era.

Later, we traveled to Prague where we listened to stories of the 1969 Russian invasion, only two years earlier. Prague was a beautiful, if bleak, city, and we soon found a café. My West German friends and I seated ourselves at a table next to a group of East Germans celebrating their vacation. After a few beers, eyes drifted from one table to the other. Finally, someone pulled the tables together, and both easterners and westerners burst into the same old German folk song. It was a beautiful sight, and it was the first opportunity that my West German friends had to meet someone from East Germany. How easy it would be to bring opposing peoples together if only the opportunities and places existed. In the early 1970s, East was separated from West, men from women, and scientists from nonscientists. Though I wasn't then aware of it, all of these fixed roles and boundaries would soon collapse.

By the time I returned to the United States, I had acquired a command of the German language, a cursory knowledge of almost every country in Western Europe, an awareness of the enormous sadness of history, and the self-assurance that comes from spending time away from the normal American college life. I didn't know yet how these cumulative experiences would help shape my life as a woman and my dreams for the future. At the end of the school year, I flew back to Michigan, to the secure lakeside world of my summertime youth.

4

SURF CITY

And not by eastern windows only,
When daylight comes, comes in the light,
In the front the sun climbs slow, how slowly,
But westward look! The land is bright

—ARTHUR HUGH CLOUGH, "SAY NOT
THE STRUGGLE NOUGHT AVAILETH," 1855

That summer, I idled in the warmth of family and soaked up the peacefulness of life at the lake in Michigan. September arrived, and with it, the beginning of my senior year in college. I felt like a new person, thrilled with political awareness. Although I had never before demonstrated for or against anything, I rapidly became involved in a campaign to save my German professor from tenure denial. As far as I could tell, the university disapproved of his friendship with a radical mathematics professor who was visibly active in civil rights demonstrations. Our save "Ottmar Jonas" campaign was fresh and aggressive. After two semesters we won, and our professor was granted tenure. That victory was sweeter to me than any single A grade or award I received.

By the middle of the year I had to face my future. Where would I go? I felt no affinity with the geologists. I felt more akin to the physicists and botanists, but I couldn't change my major now. I was to graduate in four months, and I was tired of academia, but the reality was that without an advanced degree, a Ph.D., I would spend my time in a laboratory doing little more than washing bottles. It wasn't an easy inner reconciliation, but I decided to return to my first intention: oceanography. I applied to the best schools. For once, I was aiming for the top, and it terrified me. What if I was not good enough to get accepted? One by one, the answers arrived: Yale University, accepted; University of Miami, accepted; Woods Hole Oceanographic Institute, alternate; Scripps Institution of Oceanography, no answer. Woods Hole and Scripps were the top two institutions and my top two choices. If I was going to pursue oceanography, I wanted to attend the best, and so I waited several months (which seemed like years). It was nerve-wracking, but finally Scripps's answer arrived: accepted.

At the time, my sister Ann was enrolled as a student at the University of California–San Diego, of which Scripps was part, so I visited her there during the late spring of 1973. After the lush green of Oregon, San Diego looked like it had been scorched by a bomb; tawny grasses abutted the white sand sloping down to the blue sea. Airplanes would land on the populated beaches, and recreational drug use was out in the open and everywhere. In 1973 the wave of drug experimentation somehow had passed right by me. This was San Diego: on the one hand, extremely conservative and military, on the other, free-form and hippie. I couldn't imagine where or how I might fit in.

I remember walking down the steep La Jolla road to Scripps to find Wolf Berger, one of Scripps's oceanographers, for an interview. People kept walking in and out of his office in bathing suits, carrying flippers and surfboards. These same people gathered around to discuss an upcoming expedition to the South Pacific. The expedition had a problem: They would not have enough watch standers (scientists need to monitor all the equipment for four hours at a stretch, twenty-four hours a day). "Want to meet the ship in Acapulco and ride back to San Diego?" asked one of the group. The invitation was astonishing, although I would not make my first onboard expedition until I was well into my graduate studies. This was nothing like the musty halls of university life that I had experienced up until then as an undergraduate. No one mentioned course work. Scripps was completely expedition oriented. Students wound up writing their own books because none existed that provided the necessary courses of study. Scripps was hanging ten on a big ocean surfboard.

I left the interview totally amazed, instantly dreaming of years of research in the South Pacific, swimming in balmy seas. The romance lured me like fish bait. I could feel the tangy sea air and imagine the shadows of palm fronds across the warm crystal sea. I could see freedom just out on the horizon. I signed on eagerly to life in burnt out, drug-ridden La Jolla, gleaming with new prospects.

First, however, I had to graduate. A final requirement was field camp, but geology field camp was forbidden to female students at Oregon State University. The only field camp that would accept me that summer was a marine station. I spent the summer of 1973 in St. Croix, in the U.S. Virgin Islands, working in the West Indies Laboratory. There, I obtained my field credentials by diving into coral reefs and mapping their milky blue and radiant pink branches. My job was also to take samples of the coral sediment in the lagoon by banging simple hollow pipes (cores) into the seafloor below. My colleagues and I procured our own funny research vessel, commandeered from an abandoned conch-strewn beach. A child probably built it. Its shape was triangular, and it barely stayed afloat. Nevertheless, this odd piece of marine technology carried our diving and coring equipment back and forth under the blistering sky. Its previous owner named the boat appropriately *The Arty Farty Arc,* and the letters were scrawled across the dark prow.

At the time I was studying in the Caribbean, I had only the barest suspicion that oceanography was anything but romance. I also had no idea about the vast sum of money that was invested in deep-ocean ships and equipment or that these funds were often tied to the Cold War world of stealth oceanography. Oceanography is also a tough, brutal world where its practitioners compete like pirates for ship time, data, and money, and the money needed to pay for big ships is big—up to $50,000 a day. This kind of competition breeds tough people.

Until the late 1970s, oceanographers (mostly men) were either romantics or former members of the military caught up in the serious study of the Cold War seas. Some of my colleagues were those who hungered to see the world, to explore its farthest reaches. There was wildness in their eyes, and they had little interest in academic pursuits. They preferred exploration, and the salt air and sea seemed to inhabit their bones. I was one of those romantics and as such would be one of the very few women to attend Scripps in the fall of 1973. There were so few women at Scripps that we were called "The New Girls in School" after the Beach Boys' hit single.

My sister Ann and I drove to San Diego that fall. Everything in California seemed so out of proportion. Interstate Highway 5 was sixteen lanes wide. San Diego seemed to have the most of everything: a real-life surf city, marijuana, the bluest ocean, the best burritos and quesadillas. The Mexican food was better than in Mexico, and the sun shone all year long. The cactus I brought from Oregon, once transplanted, lifted its head, its spines stiffened, and the green fibers grew taut. This was a different world. Ann and I shared a house with a group of her friends in Cardiff, up the street from V.G. Donuts, the famous surfer hangout. From Leucadia, Encinitas, Cardiff, Del Mar to La Jolla, the beach was crammed with surfers, and the adjacent roads were bumper to bumper with painted Volkswagen vans. A ratty bus called the Coast Cruiser plied Route 101, picking up and dropping off students.

We shared an old blue Ford station wagon nicknamed "the boat." Every morning, two of our uncommunicative housemates, my sister, and I drove to La Jolla. After we parked, I walked down the long hill to Scripps, past farms, past the enclaves of Scripps professors, past the fisheries building perched over the sea, past the Institute of Geophysics, past the groves of Torrey pines, past the brilliant velvet purple ice plants that carpeted the steep cliffs, and past the red flame trees from South Africa. Finally, I approached Sverdrup and Ritter Halls down by the sea, the main campus of Scripps Institution of Oceanography. Every day, I walked from the hippie world to the confusing mix of the pirate and military world of oceanography. Perhaps that contrast explained why society and the science of oceanography changed so radically in the 1970s. Those of us living through that change were irrevocably affected, especially by the eventual though incomplete acceptance of women in science and, later, the slow exit of the military from the science of oceanography.

That first day in Ritter Hall, I met Betty Stoffer, the Scripps student secretary. We arranged classes—it was easy; everyone took the same courses—and then I met a fellow student, Larry Mayer from Rhode Island. There were fewer than fifteen new graduate students in the class of 1973, and within that group, only four geologists. In addition to the Larry Mayer from Rhode Island, we had another Larry Mayer, from Maine, whom we distinguished by his straight hair. Other friends included Jan Hilson from Michigan,

Marcia McNutt, a geophysicist from Colorado, Russ McDuff, a chemist from the California Institute of Technology, and Ted Dengler, a biologist from the Massachusetts Institute of Technology.

The "core courses" we took comprised physical oceanography, marine biology, and marine chemistry. We geologists added to the core a marine geology seminar, a course in geochemistry, and another in geophysical methods, in other words, "what ships do at sea," whether it was booming out sound by detonating charges or listening for secrets from the abyss. I remember great success with physical oceanography, loving the history of the sea and learning about its vast currents. Biologists, chemists, geologists, and engineers took these courses. None of us had much background in the actual science of the ocean, as we had been trained only in the classical sciences.

Oceanography was entirely new, and therefore there was a steep learning curve. The field was interdisciplinary in nature; the biology of the sea affected the chemistry, the chemistry affected the geology of the seafloor, and the motion of the deep currents affected everything. To study the oceans was to study a holistic world of interdependence. Our role was to integrate our specialties with all the other areas of the ocean sciences. I was fascinated when I learned about the deepwater currents that had just been discovered (the North Atlantic Deep Water and the Antarctic Bottom Water). The massive National Science Foundation program to unravel the geochemistry of the sea by sections (GEOSECS) had just been completed, and chemical tracers had started to reveal magnificent stories about these immense deep-sea currents. For example, oceanographers had learned that water sinks to the seafloor in the Arctic and the Antarctic, hugs the continental slope, and cascades around the world carrying oxygen and nutrients in its wake.

Scripps housed world-class scientists, many of whom were marvelous characters. Some were competitive, others vindictive, but all were colorful. The chemistry department was particularly difficult; it was filled with high-profile researchers, some of whom disliked one another. Harmon Craig, who traced helium-3 in the ocean to produce evidence of gas emissions from deep in the earth's mantle, never worked during the day, but only between the hours of 6 P.M. and 6 A.M. His domain, which included a bright graduate student, Ray Weiss, and Craig's wife, Valerie (she monitored the gates to Craig's research fortress and expelled any unwanted visitors), was world renowned both for producing brilliant pieces of scientific research and for its difficult personalities. The leader of the opposing camp was the equally famous Ed Goldberg who, of course, never worked at night. The unwitting graduate student caught between the two fiefdoms was torn to shreds like shark bait.

In marine geology were Wolfgang Berger and Jerry Winterer. Winterer could always be counted on for his wonderful stories of the South Pacific. He could be full of warmth and camaraderie one moment, and in the next, he would burst into anger that would send people running. Fred Spiess, a former submarine commander and physicist, ran the famous Deep-Tow Group, a tough outfit he commanded as if he were still in the navy. In addition to the soft-spoken Spiess, whom everyone called "God," Tony Boegeman, an engineer, ran the group as if he were marshalling troops for war. The young Vince Pavlicek was the group's new engineer. John Mudie, a brilliant South African geophysicist, worked with magnetics, calculated cable trajectories, and produced software that ran the Deep-Tow sonar. Steve Miller, who came off like a smart hippie, was the group's computer operator, and he mingled seamlessly with the more military types. The group was rounded out by an upstart group of pirate-like graduate students who angled to run the program. Deep-Tow students were known for becoming chief scientists by their third year in graduate school. These were not soft people.

Merle Henderschott, a hotshot mathematician, possessed an unbelievable talent with the pipe organ. Merle told a story about the tidal wave (tsunami) that was supposed to smash into San Diego some hours after the 1964 Good Friday earthquake in Alaska. According to Henderschott, local surfers ran to the beach looking for the "great wave," while all the oceanographers packed their belongings and ran to the highest ground. Henderschott left his data behind but hoisted up his collection of Bach recordings and brought it with him. The great wave never hit, yet even today the surfers are still out on Scripps Beach waiting for "the Big One."

In the bright autumn, blue light streamed inside the Scripps buildings, bathing the white stucco walls of the offices. The most sought after oceanographer, Bill Menard, inhabited a corner of Ritter Hall overlooking a broad blue expanse of the Pacific. He used to write on a blackboard his most important scientific questions: "What forms seamounts? What are oceanic swells? How do fracture zones develop?" To the left of his famous blackboard, tiki dolls lined the bookcases or were scattered over the floor, depending on whether or not an earthquake had shaken Del Mar the night before. Large seashells from the South Pacific—orange-splotched cream and ochre turbots, trumpets, and cones—lined the crevices between the hundreds of books. Papers were stacked high everywhere. Old and unfinished manuscripts filled his desk together with a jumble of notes, odd jingles, and poems. Menard was the most romantic of the "old school" oceanographers. He practically invented our present-day concept of the Pacific seafloor. He never stopped asking questions about its hidden mysteries. He was the first to discover the connection between seamounts and seafloor swells, the first to lead expeditions to the South Pacific, and the first to write it all down. His office was warm, full of life, and full of questions about the earth.

My first year at Scripps exposed me to these incredible people. I was privileged to know Menard and Russ Raitt, a geophysicist who threw the most phenomenal parties in his open house perched above the Pacific. Mexican food started at 6 P.M., and dancing to music provided by Raitt's niece Bonnie went till dawn. Raitt's wife, Helen, specialized in Polynesian lore. The house was filled with canoes, shells, bamboo, echoes from the South Pacific.

George Shor, another geophysicist, collected rare fruit trees and plants from deserted islands and brought them back to San Diego aboard Scripps's ships. Once I heard that he had strapped a twenty-foot cactus to the ship's mast in order to give it light as they returned north from Mexico.

My first year at Scripps was also a journey among the colorful animosities and intense egos that inhabited its halls. During the 1970s, "big science" and the military tangled behind the laid-back surfer lifestyle. Ironically, the rivalry was less intense if a professor was heavily involved with the military, because the military represented money, and when money was at stake, the fight for recognition was less important. Scientists grabbed data, locked it away, openly derided the nearest competitor, and eliminated those graduate students who were not aggressive enough. Too often, female students were shut out from the decision-making processes entirely—unless they assumed an even more aggressive demeanor than their male colleagues. A tough outer shell was essential, if a woman wanted to survive.

Many of the male students had just returned from the war in Vietnam. Most were older students with wives and families at home, and their families supported them financially as well as emotionally. The female students, on the other hand, were a new breed: younger, single, and forced to rely on their own resources to make it through graduate school. Even though UCSD's undergraduate school was very liberal in the early 1970s, Scripps remained conservative. Enough change had occurred so that it had opened its doors to women, but it was also understood that if a woman wanted to be a scientist, she would generally have to choose between a personal life and a career. Out of the seven women who were graduate students with me, only one has ever married.

At the same time, interspersed in all of the nastiness, intense creativity and excitement abounded. As students, we were challenged to question existing theories, to make great discoveries, and to push our emotions and insecurities into the deepest recesses. It was plain: Fight and be right; defer and be thrown out. Demand and lead, or be rejected. Thrown into the mêlée, where scientist-gladiators surfed the ocean during lunchtime, we students rapidly embraced the fight. The once shy but strong girl of my youth was now determined to succeed.

My first research experience would be with the Deep Sea Drilling Project (DSDP), an international program with the goal of drilling into the ocean's sediments and crust to test the various theories of plate tectonics. The results from DSDP showed that the present-day oceans were very young compared to the buoyant continents. The research also indicated that the history of global climate change could be deciphered from the deep oozes of the abyssal seafloor. The DSDP got its start in the early 1960s as a spin-off from ideas spawned by tipsy Scripps professors attending a champagne brunch when President John F. Kennedy announced that America would put a man on the moon by 1969. The professors wondered, "Why don't we drill a hole through the Moho boundary into the earth's mantle, through the ocean crust?" Shortly thereafter, Project MOHOLE was born (named after Andrij Mohorovocic, a Croatian-Yugoslavian seismologist who discovered the transition that separates the earth's topmost layer of crust from its upper mantle). Ten years later, Project MOHOLE still had not succeeded. Instead, it had grown and diversified along more traditional lines of scientific inquiry. The thrill-seeking, goal-oriented, entrepreneurial oceanographers had long ago bailed out.

My salary during that first year was paid by DSDP—that is, until I changed course and cast my lot with the tough Deep-Tow Group.

5

THE DEEP-TOW GROUP

The first beginnings of wisdom is to ask questions
But never to answer any.
—FLANN O'BRIEN, *THE THIRD POLICEMAN*

Fred Spiess, director of Scripps's Deep-Tow Group, commanded more than respect among the echelons of his staff. The uppermost levels seemed to be ex-military personnel. Spiess certainly was—he had commanded submarines in World War II. His word was revered. Everyone was frightened of his power, and though I never saw him abuse that power, I surely quivered at his feet. Scripps was chock-full of Ph.D.s, but there was only one "Doctor," and that was Dr. Spiess. People with whom he was just casually acquainted called him "Fred," and only Sally, his amazing wife, called him Noel, his middle name. He will always be "Doctor Spiess" to me.

Sometimes it took me weeks to screw up the courage to schedule a meeting with him, and I did so no more than once a year. Dr. Spiess told me, "Kathy, I don't care where in the world you do your work just as long as you get it done." That was good enough for me.

Perhaps Dr. Spiess seemed so aloof because it was the early 1970s and he'd never had any experience with female graduate students. To me, he was a navy man through and through. To be a woman and to march into his camp would have been brash, if not downright stupid. Nevertheless, I saw in the Deep-Tow Group an opportunity to go to sea, to establish new research programs, and to direct them. Oddly enough, although tough and disciplined, the group was not subject to the passions of many of Scripps's other laboratory divisions. The Deep-Tow Group belonged to the Marine Physical Laboratory, which directed all the navy-related acoustics experiments. The lab's former name was the Scripps Laboratory for Naval War Research, a name that didn't wash with the student radicals of the 1960s. The name of the lab had changed, but its mission had remained the same. Classified research was still conducted at the Point Loma Naval Base, where the Deep-Tow Group kept its navy-related projects under wraps.

Once, my sister Ann, a reporter for the activist newspaper *The Natty Dred,* tried to investigate the classified activities of the Deep-Tow Point Loma naval facilities. I later heard through a colleague that she had arranged an interview with Spiess's chief assistant. According to my colleague, the assistant looked at my sister and said, "Say, you look just like one of our graduate students." The interview stopped when Ann realized that the assistant was referring to me and that I worked in a place that displayed signs that said "Loose Lips Sink Ships."

I was proud of my sister for attempting to discover what went on at Point Loma, for in truth, I, too, had no idea. Graduate students were not allowed to work on classified research, and that suited me fine. Nonetheless, other students at the University of California were very suspicious of our work in the Deep-Tow Group. On one occasion, two students stopped Larry Mayer and me while we were walking on the upper campus. "Hey, don't you guys build bombs?" they asked, pointing to our badges. We just stared at them without knowing what to say.

The kind of work the Deep-Tow Group did involved data collection and analysis. The Deep-Tow was the first instrument of its kind: a side-looking sonar (also called a side-scan sonar) that was lowered into the water off the fantail (or back end) of the ship. It was literally towed in the deep ocean as its instruments recorded sonar images, which looked like sound-based photos of the seafloor below. Dr. Spiess had developed it. The navy had an interest in it because it could map the location of sunken submarines. In fact, the Deep-Tow was first used to search for the USS *Thresher*, the nuclear submarine that sank off Cape Cod in the 1960s. With the addition of a magnetometer, quantitative data could be gathered about the nature of sunken craft, as well as the spreading rates of mid-ocean ridges, those places on the seafloor where two plates are spreading apart. Metals are highly magnetic, and a magnetometer was the ideal instrument to enhance Deep-Tow's capabilities to search, locate, and identify.

A year and a half passed, after which Larry Mayer and I inherited our own office from fellow student Kim Klitgord after he graduated. Our office was in T-5, a wooden house by the shore. The "T" in T-5 stood for "temporary structure," built in World War II, and it housed many graduate students, one or two per room. Peter Lonsdale camped out in the next room, Bob Detrick had the back office, and Jim Natland, known as "The Nat," sat next to Peter. All three were a couple of years ahead of me in graduate school, and they knew all the ropes. Peter piled charts everywhere and crammed each corner with sonobuoy cases. (Sonobuoys are a type of equipment that automatically transmits a radio signal when triggered by an underwater sound. They are often used to monitor deep-sea earthquakes.) Sun-faded transponder flags, rock samples from the seafloor, and manganese nodules dredged up from the Hawaiian depths were used as doorstops. The ancient, rusty sink was filled with dark coffee cups and drying bathing suits, while surfboards leaned against the walls. Rolls of 3.5, 4, and 12 kilohertz (kHz) echo-sounding records were stuffed into every nook and cranny, and hundreds of thousands of dollars worth of Deep-Tow side-scan records were lined up like huge cigars on top of the paper-strewn drafting table. T-5 may have been home to more budding prominent oceanographers than any other building in the world. Here we could be free from Dr. Spiess, from secretaries, and from phone calls. Here we dreamed up our escapes, our proposals, the direction that our science would take, and the incredible treks all over the globe on our way to and from research expeditions. We became travel experts. If we were called to work in Bali, we could rig up a route that would take us across the Atlantic to the Kalahari in Africa, across to New Delhi, and finally to Bali, and at half the normal price. We were intent on uncovering the secrets of the oceans and seeing plenty of lands on the way.

Every day, Peter tramped in at 7:30 A.M., after the long walk up the La Jolla coast from Nautilus Street. He never wore shoes, and the soles of his feet were as hard as nails. He always wore sunglasses, however—in class, at night, in restaurants, in broad daylight. Some said he wore them in bed as well. Peter was British. He would fling his brilliant sarcasm all over the place, infecting anyone it hit. He was my office neighbor.

The Nat lived with Peter in a house on Nautilus Street. The Nat was a nice, bookish fellow, and seafloor rocks meant everything to him. He lived buried in rock samples, glass slides, and microscopes. Their Nautilus Street house provided an asylum for other Scripps oceanographers as well. Steve Huestis, of hang-gliding fame, and Clark Wilson, an avid bicyclist, often camped out. Sometimes Ted Dengler, a hockey-playing marine biologist, would move in for a spell, and sometimes I would live in the laundry room after I returned from sea, with no place to go because had I rented out the room where I ordinarily lived. Sometimes Ken Macdonald would come out from Woods Hole, to escape from Cape Cod's frigid winters. He would unofficially move in to one of the other "T" buildings on campus; he would climb in the windows at night and unroll his sleeping bag. When he was caught, he too crashed in the laundry room in the house on Nautilus Street. The laundry room was a great set-up: It was warm because of the dryers, it had a toilet, a refrigerator nearby, and a quick exit to the backyard and the neighboring alley.

Nautilus Street household: Back row (left to right): Kathie Carpenter, Ouvene Klitgord, Ted Dengler, Steve Miller, Marcia McNutt, Kim Klitgord, Ruth Levy, Steve Huestis, Bob Tyce, Clark Wilson. Front row: Kathleen Crane, Jim Natland, Peter Lonsdale, and Ellen Fredricson

Peter and Steve Huestis concocted many famous and infamous Nautilus Street dinners. The only things Peter knew how to cook were sausages and plums, and when plums were in season, the dinners included "plum everything." They inaugurated the "soon-to-be-famous" Oceanographic Twinkie-Eating Contest. The top contenders included Kathie Carpenter (Marcia McNutt's roommate) and Steve. And they really did it: tequila and a Twinkie, one after another. Steve won at twenty-two rounds and shortened his life by ten years in the process.

The Nautilus Street house hopped with fantastic parties. Whenever any oceanographer came to town, it was time for burritos, margaritas, and the Rolling Stones. Steve rigged up great dance music that reverberated down to the ocean. The cakes were the crowning event of these evenings. Peter and I would drive down to Tijuana right before any party and scramble through the streets searching for the most grotesque pastries. All the cakes had themes: cowboys and indians, pirates, banditos, all in Day-Glo-colored frosting. We were not finished celebrating until our mouths were stained by a green frosting hill or a radiant blue frosting wave. Our teeth were tinted for months after these Mexican indulgences.

Steve also enticed members of our group to attempt tremendous skateboarding feats down the Scripps hill. This was generally a near-death experience for those of us who were "weenies" and had to bail out of the race at the Fisheries Institute curve. I was a classic, cowardly "weenie." Steve, in contrast, was the only one who made the entire run unscathed. Later, he took up hang gliding off cliffs and sand dunes above the ocean, and we all helped him from time to time by lugging his flying gear. Once I helped him at 6 A.M. (supposedly before the campus police made their rounds) to the roof of the Institute of Geology and Planetary Physics (IGPP), a building perched on a steep cliff above the Pacific Ocean. Just as Steve leaped off the roof and sailed out over the roiling sea below, the police showed up, and shortly thereafter, UCSD posted a notice that forbade hang gliding from its buildings. This notice did not dissuade Steve much. He wrapped up his career at Scripps with a funny opening to his thesis defense, where he quipped, "I hope you realize that this is really cutting into my flying time."

These adventures punctuated our research schedule, outrageous stunts in between nights of working on the computer or grinding out numbers at Scripps. Computer rates were cheaper at night, so the really intense students worked between midnight and 6 A.M.

Steve, Clark Wilson, and Kim Klitgord worked together on geophysical problems related to the variations in Earth's magnetic or gravitational field. Kim collected real data at sea, while Steve and Clark manipulated "ideal data sets" or "ideal bodies," as they called them. Seafloor rocks emit a magnetic signature because they are often made up of basalt (volcanic rock that has cooled and solidified over time), and basalt is filled with magnetic minerals. When hot basalt cools to about 1072° Fahrenheit (578° Celsius), the minerals, like tiny compasses, align themselves in the direction of Earth's magnetic field. Today, the magnetic lines of force around Earth enter the planet at the North Pole and exit from the South Pole. At other times in Earth's deep past, the magnetic lines of force were reversed. We know today that Earth's polarity has switched over time—the present North Pole was once the South Pole and vice versa.

Because most of the basalt forms at mid-ocean ridges, it is rafted off the ridges as Earth's tectonic plates pull apart and separate. As new crust is minted and moves away from the mid-ocean ridge, new magma is erupted at the center of the ridge.

The movement of the plates acts like a magnetic tape recorder of Earth's changing magnetic field. When the magnetic poles of Earth reverse, new iron-rich magma is imprinted with a reversely oriented magnetic signal. This process is repeated endlessly over great spans of time. The result is magnetic striping of the seafloor. Oceanographers have been able to determine when Earth's magnetic field has flipped and can calculate the rate that Earth's plates are moving—the rate of seafloor spreading. In using our tools to decode this signal, we needed to position the Deep-Tow instrument and its attached magnetometer as close to the seafloor as possible. This resulted in a much higher data resolution than if we had towed the instrument at the sea surface, several miles above the seafloor.

Klitgord and Huestis attempted to decipher magnetic anomaly patterns across mid-ocean ridges located close to the present-day Equator. Theoretically, basalt that forms and cools midway between Earth's magnetic poles is imprinted with only very weak magnetic signals. However, no one had actually looked closely at the magnetic patterns near the Equator. Therefore, Klitgord concentrated his efforts on the Galápagos Spreading Center (the East-West oriented mid-ocean ridge that straddles the Equator in the Eastern Pacific). Scientists working at Woods Hole Oceanographic Institution later called this ridge the Galápagos Rift. Klitgord's interest set the stage for all of Scripps's later work on this mid-ocean ridge, which culminated in the Deep-Tow discovery of hydrothermal vents there in 1976.

The Deep-Tow images of the seafloor were produced by a unique set of transducers, electronic instruments that send out beams of sound at a high frequency (115 kHz) on each side of the "fish" (which is what we call any object towed in the deep ocean behind a ship). In this case, the fish was the Deep-Tow. These sonar signals, acting somewhat like beams of light, hit the objects on the seafloor, "illuminate" them, then are reflected back to the fish. The amount of sound that the seafloor reflects depends upon the nature of the material on the seafloor. Fresh basalt from recent lava flows acts almost like a mirror and reflects sound nearly perfectly. Most of the acoustic signal hitting basalt bounces right back to the fish, where it leaves a dark trace on a recording device, which indicates a strong reflectivity. Steep topographic features, like volcanoes, also reflect sound efficiently. In contrast, sediments absorb the sound signal, and only weak traces make it back to the fish. The amalgam of these acoustic signals produces images of the seafloor, which look similar to photographs taken of Earth from an airplane.

Over time, the original Deep-Tow device evolved into a fantastic contraption that housed almost any kind of instrument that a graduate student could imagine. Larry Mayer developed a pump and a suspended sediment sampler, which, when attached to the Deep-Tow, made it look like it had giant ears. First Bob Detrick and then I developed a temperature-monitoring system that dangled below the Deep-Tow and was used to search for hydrothermal activity on the seafloor. Karen Wishner, the first biologist to use the Deep-Tow, invented a set of plankton nets that were electronically controlled and could open and close on command. These were stuck onto the bottom of the Deep-Tow, making it look like a head with flapping jaws. Many other devices were attached from time to time to the side of the Deep-Tow. It also housed the first still-frame deep-sea video system so that we could film in real time. It was always an exciting event when we lowered the Deep-Tow to within only a few feet of the seafloor to photograph fissures, faults, and lava flows. At the same time, we could take temperature readings of the bottom water, and if we noted a suspiciously warm reading, we could open and close a sampling bottle using remote-control commands.

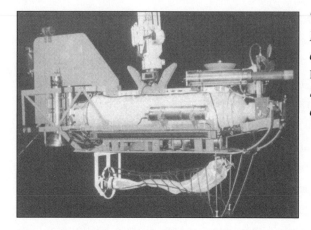

*The Scripps
Deep-Tow,
circa 1976*
PHOTO: *Phil Stotts,
Scripps Institution
of Oceanography*

Satellites and dead reckoning were used to navigate the ship that towed the Deep-Tow. Once the Deep-Tow was deployed, it was navigated by using a system of bottom-anchored transponders that communicated at low frequencies of 10 to 11 kilohertz (kHz) with both the Deep-Tow and the ship. This type of bottom navigation system was so precise that we could maneuver our equipment with the greatest accuracy ever achieved in the deep ocean. Before the development of these transponders, navigation was so imprecise that oceanographers were unable to return to the same site in following years. In earlier times, some oceanographers reportedly welcomed such imprecision. Their discoveries never would be challenged because their survey sites could never be relocated. In the early 1970s, depending upon location on Earth, satellite fixes could be made only a handful of times each day. Today, positions can be determined within 100 meters every few seconds.

Our cutting-edge technology in the 1970s made us the "SWAT team" of the deep. Nothing was out of bounds; we could search anywhere and map any place. Our instruments were among the first to test hypotheses about the fine-scale evolution of the mid-ocean ridge, sediment transport, the distribution of animal life in the deepwater column, and the interrelated nature of deep-ocean chemistry, physical oceanography, and the geology of the seafloor.

The Deep-Tow required a team of four to stand watch because operations near the seafloor demanded constant attention. One person (the flyer) flew the fish by raising and lowering the cable, one navigated around and through the transponder net, one was the watch leader, and one took notes. The ship's engineers and computer specialists were "floaters" and were called in when problems occurred with the cable, when there were near-bottom camera runs, and during launches and recoveries. Each watch was hectic. Because the fish was flown within 50 fathoms (300 feet) of the seafloor, the fish flyer would keep constant watch for rapidly approaching volcanic peaks or fault scarps (which appeared as sudden, sharp cliffs), which could snag the delicate instrument. Sometimes the optical avoidance sonar revealed these kinds of seafloor features too late, and our only recourse was to manually hoist the cable of the Deep-Tow and to accelerate our speed to 10 knots (10 nautical miles per hour). The ship normally maintained a speed of 1.5 knots so that we could bring the instruments close to the seafloor. Accelerating the ship's speed lifted the fish away from the seafloor and away from danger. When we were off watch, we could tell that the watch leader was having problems by the sudden high pitch of the winch that was pulling in cable and the abrupt change in engine speed. We worried constantly about the possibility of losing the Deep-Tow to volcanoes below, mostly because we feared the wrath of Dr. Spiess.

The Deep-Tow Group's watch was very demanding. Those who stood watch did so for four-hour stretches, twice a day. Generally, the most senior people took the 4–8 watch, the sunrise and sunset watches, the least experienced took the 8–12 watch, and the "middle management" took the 12–4, the "dead" watch, or "grave-yard" shift. Outside of watch change, team members rarely saw one another until the ship steamed into port at the end of a cruise. On some expeditions, the watch leader wore a cap and passed it on at the end of each watch. The watch team would have to show up ten minutes before the hour to be briefed on what happened during the last watch and to receive the expedition's chief scientist's instructions for the day.

The watch leader monitored the seafloor, searching for approaching volcanoes and faults by checking the side-scan sonar images. The leader also called up to the bridge crew who were steering the ship to notify them if a change in the course or the heading of the ship was needed. These course changes often took place five degrees at a time, many times an hour, in order to keep the Deep-Tow on its predetermined track, which we called "the yellow brick road." At the same time, the navigator used three differently colored pens to rapidly trace the slant ranges (the line of sight distances between two points not at the same elevation) from the transponders to the fish. Three frequency signals or traces—10, 10.5, and 11 kilohertz, also called green, red, and blue, respectively—coursed across the echosounder. Sometimes the navigator had to track up to nine traces at varying degrees of visibility and ranges on a five-minute basis. Precise navigation was crucial, and one-minute fixes were ordered so that the watch leader could make rapid course changes through difficult terrain. The navigator's job was the most demanding because he could never relax, never slow down for a minute, over a period of several hours. After a watch, the navigator's hands were striped with green, red, and blue, and wild-colored traces splashed across the recorder paper. At that point, the only option for the totally exhausted navigator was the bunk.

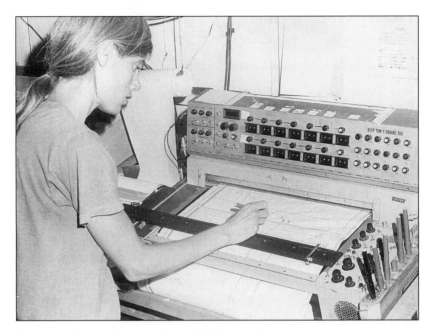

Transponder navigator. Photo: *Phil Stotts, Scripps*

Every watch moved like clockwork. People arrived, messages were transferred, information was entered into the logbook, and every system was double-checked. A typical watch book contained information as diverse as "Zulu" time (Greenwich Mean Time), course, wind speed, sea conditions, changes in course and speed, satellite fixes, wire out, winch problems, and the occasional personal observation. Entries varied wildly:

> 1135 Scale change on 3.5 and 12 kHz echosounders to 200–600. At approximately 1715 last charge will go over. Contact bridge to cease radio transmission for approximately 5 min.
>
> 1200 Donovan's thinking of starting a meat market during Lent.
>
> 1530 c/c 171 degrees 2045 ETA at PT.A.

1703 c/c 165 degrees—We picked up the counter-current!

1732 NOTICE: underway watch people and everyone else. Throw your own trash away or else someone after you will throw you away!

When the Deep-Tow was recovered from the sea, every member of the team was called to action. First, Dr. Spiess would take the commander's stand on the fantail, and with arms raised and fist clenched, signaled the Deep-Tow crane operator to swing the fish over and onto the deck while keeping the cable taut. The geometry of the cable layout was very complex. From the fantail to amidships, the cable angled horizontally back and forth across the deck until it was routed vertically to a drum in the hold of the ship. Students were directed to pull on the cable to remove any slack during launch and recovery.

Pulling the Deep-Tow cable. Left to right: Dr. Spiess, Bill Normark, John Shih, Steve Miller, and Vince Pavlicek

The name of each Deep-Tow cruise ended with "tow." For exam-ple, the *Southtow* expedition was named for work conducted in the South Pacific, the *Tic-Tac-Tow* expedition for a grid survey for the navy, the *NATOW* expedition for a North Atlantic tow, and the *Coco-tow* expedition for surveys of the Cocos Plate near Central America. The classified search for a sunken Russian submarine off Hawaii was named *Missiletow* (but no graduate students were allowed on that trip). There were endless possibilities: Broken Tow, Hurt Tow, Sprained Tow, Row Tow, Oh No Tow, and No Hope Tow.

Soon, I would head off on my first expedition.

6

A CONUNDRUM

Where observation is concerned, chance favors only the
prepared mind.

—Louis Pasteur, address given on
the inauguration of the Faculty of Science,
University of Lille, December 7, 1854

As I approached the end of my first year at Scripps, I needed to
define the topic of my thesis. The exotic deep sea was luring
me, and it was the mapping of volcanic activity erupting out of the
mid-ocean ridge that I finally settled on as my focus of study.

In the early 1960s, marine geologists thought that the amount of
heat emanating from Earth's mantle through the seafloor would be
highest at the volcanic mid-ocean ridges. But as geophysicists pushed
their heat flow probes into the sediments of the ocean floor, they
found that the conductive heat flow was much lower near the mid-
ocean ridge than they had expected. Conduction, or heat transfer
without the movement of material, was the only type of heat flow
measurable at the time. The "missing" heat baffled geophysicists
until a few of them guessed that the heat must have been escaping
into the ocean through the eruption of hot water springs. When
heat flows with moving material, such as hot water or magma, the
process is called convection. Because the seafloor near the mid-ocean
ridge is dotted with underwater volcanoes, hot water in the oceans

above hot springs should be abundant (heat + water should equal hot water). But no one had ever seen a deep-sea hot spring before, and no one had conclusive evidence that these "vents" into the sea might indeed exist.

In fact, because of the great pressure on the seafloor, there was some speculation that hot springs could not form. In addition, water cannot boil unless temperatures are extremely high. During the 1970s, very little was known about the chemistry of saltwater. However, the largest factors hindering the discovery of hot springs on the deep-sea floor were the lack of available technology and the conservative nature of scientists, which created a psychological barrier against exploration.

Since that time, many books have been written about the discovery of hydrothermal (hot water) activity on the deep-sea floor. Many of these attribute the discovery to the first visit by humans to the Galápagos Spreading Center in the submersible vessel *Alvin* in 1977. However, the discovery of hydrothermal activity on the seafloor has a longer history, dating back to the 1880s when hot water was discovered in the Red Sea by the Russian research vessel *Vitaz*. The Swedish ship *Albatross* confirmed this astonishing finding in the 1940s. However, even though Jules Verne wrote about undersea volcanic activity in *Twenty-Thousand Leagues Under the Sea*, oceanographers were very slow to accept the idea that hot vents really existed. Instead, renowned oceanographers preferred the idea that the hot, briny, and heavy water formed at the sea surface under the heat of the tropical sun and sank to the seafloor below.

Another twenty years passed, including World War II, before new discoveries were made, again in the Red Sea. Three more ships, the *Discovery* from Great Britain and the *Atlantis II* and the *Chain* from Woods Hole Oceanographic Institution, all confirmed that high-temperature (44°–56°C) briny waters and metal-rich mud were baking and oozing in the sea's deep basins. However, due to increasing political unrest and war, the Red Sea soon was designated off-limits to research vessels and further exploration. At that time, Egon Degans and David Ross, scientists from Woods Hole, proposed that other mid-ocean ridges across the globe should also be sites of deep-sea hot springs and heavy brine activity.

Luckily, it was found that large quantities of metal-rich sediments draped the flanks of the mid-ocean ridge in another ocean, practically on the opposite side of the earth from the Red Sea, and yet these sediments were very similar in composition to those in the Red Sea. While politics was shutting out the free passage of ships to the Middle East, all oceanographic eyes turned elsewhere—to the Pacific Ocean.

Although the proposal of exploring for deep-sea hot springs seemed like a simple one, securing research funds to find them was a formidable challenge. No one had mapped the temperature of the ocean bottom water with any regularity. In the past, when temperature anomalies had been reported, most physical oceanographers ascribed the data to erroneous noise. Yet this is what we had to find: the missing convective heat that the existing conductive heat flow instruments could not measure.

Signs in the deep Pacific Ocean suggested that indeed a gigantic source of heat warmed the bottom waters above the mid-ocean ridge. Over much of the East Pacific Rise, isotherms (lines of equal temperature in the ocean) bent down toward the mid-ocean ridge. Normally, isotherms are warmer toward the sea surface, because the water there is heated by the sun, and much colder in the deep, averaging temperatures barely above freezing. Isotherms also tend to be horizontally stratified; that is, they appear to be layered, one on top of the other. Scripps's physical oceanographers explained that the (layers of) isotherms bent down toward the East Pacific Rise because of the drag of the bottom currents across the high sections of the mid-ocean ridge topography. A few others, including myself, thought that the isotherms bent down due to tremendous heating from below.

Other senior scientists agreed with the possibility but were not committed to researching the idea. My first thesis advisor, John Mudie, a Deep-Tow geophysicist, was skeptical. When he reviewed my proposal to head out with the Deep-Tow Group to search for underwater hot springs, he asked, "What happens if you don't find any?" This made me nervous. A well-established scientist at Woods Hole added, "You'd better find another thesis topic. You'll never find hot springs on the seafloor." I nearly grew dispirited.

I was surprised to find that Dr. Spiess supported the idea. Bob Detrick, then a student in the Deep-Tow Group, had been calibrating the Deep-Tow oscillating quartz thermometer on a regular basis. He had participated in the 1972 *Southtow* expedition to the Galápagos Spreading Center, and it was there that he and fellow scientist Dave Williams found temperatures in the water column that were several hundredths of a degree warmer than the surrounding water. There were other signs of volcanic activity at the Galápagos Spreading Center. On another *Southtow* expedition, Ken Macdonald and John Mudie were monitoring earthquakes when suddenly a large number of dead deep-ocean fish rose rapidly to the surface. The fish were sampled and taken back to Scripps where they were confirmed as "benthic" (deep bottom dwellers). Something big must have happened to have killed so many fish, and Mudie and Macdonald speculated that a deep-ocean volcanic event must have destroyed this community of fish.

These results excited me, but Bob Detrick warned that there might be some problems with the thermometer system and added that he wasn't sure how significant the tiny temperature anomalies were. Still, I was determined to take a new approach to mapping the bottom water as I searched for the elusive hot springs. I would conduct the mapping of temperatures in three dimensions over a section of the mid-ocean ridge. Dr. Spiess allowed me to lower the quartz thermometer on a cable underneath the Deep-Tow so that it could take measurements closer to the seafloor, closer to the source of volcanic heat. Vince Pavlicek, the new engineer in the Marine Physical Lab, was assigned this task. We were both twenty-two years old.

My goal was to test the new thermometer during the upcoming expedition to the East Pacific Rise. This would be my first real cruise, my first chance to the see the Pacific Ocean, and my first opportunity to seek out underwater hot springs. Little did I know that it would be a journey into hell.

7

THE FIRST SEARCH

A journey is a hallucination.

—FLANN O'BRIEN

I t was the summer of 1974, and the Deep Tow was being readied for exploration of the East Pacific Rise (near 21° North followed by 8° North where it intersects the giant Siqueiros Fracture Zone). A small part of our exploration was dedicated to the search for the elusive underwater hot springs. Bob Detrick was supposed to be the co-chief scientist for the *Cocotow* expedition along with John Mudie. As the departure date drew near, however, strange things began to happen to the assembled crew. Without any warning, Detrick announced that he was leaving graduate school. John Mudie was reported to be experiencing personal problems. In the 1970s, it was common practice to seek out alternate therapies, and Mudie was no exception. Under intense stress, he decided to completely revise the shipboard roster, adding several people who were not oceanographers but people he had met during "encounter" sessions in San Diego. When Detrick resigned, I was the only geology graduate student left on board. Computer specialists (Ron Moe, Steve Miller, and Kathy Poole), Deep-Tow biologist, Karen Wishner (who would test her newly developed plankton net), Phil Stotts, a Deep-Tow photographer, and a host of

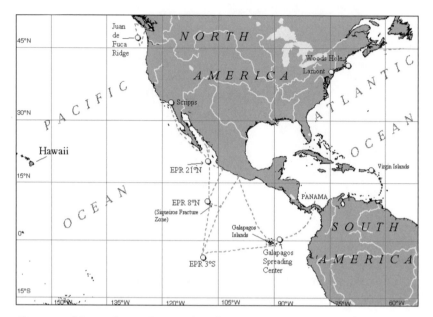

Cocotow ship tracks in Eastern Pacific

volunteers rounded out the participants. Shortly before we left port, another astonishing change occurred. Bill Normark, a marine geologist from the U.S. Geological Survey, was appointed chief scientist, replacing John Mudie during the first two weeks at sea. No explanation was given. Mudie would join us in Manzanillo for the last leg to the Siqueiros Fracture Zone. Mudie's encounter group remained on board.

Those of us in the Deep-Tow Group were skeptical. Why did the chief scientist disappear? The Deep-Tow's senior engineer, Tony Boegeman, grew so disillusioned that he refused to participate on the last leg of the expedition. Instead, he would send his junior, Vince Pavlicek, and the seasoned Martin Benson (whom we called Curly).

The crew piled onto the *Melville* in the San Diego harbor on a warm sunny day in August 1974. We fueled up at the pier and departed at 2045 Zulu (Z) on August 26 and steamed away from the marine facilities out to the blue Pacific. At 2330 Z, we received notice of engine failure and changed course to return to San Diego. We arrived back at the marine facilities for repairs and departed one more time the same day. Once out of the harbor, we changed course and headed south toward the East Pacific Rise.

This was my first long expedition. Normally, first-timers would spend the first month learning the ropes, working up from note taker on watch to navigator. I felt pressure to learn everything at once, never admitting that I didn't understand. Tony Boegeman ran the electronics racks like a Marine. These vertical panels were the nerve center for the complex Deep-Tow operations. All scientists were ordered to keep their hands off. Although I was quaking, I tried to act as tough as possible.

Before we left port, Karen Wishner and I were warned about "showing too much skin" on the ship. We were advised to wear long, loose sweaters so as not to attract attention from the regular crew members. We were lectured at length about how women in the past had acted seductively and how this had caused onboard strife. In retrospect, this treatment was wholly inappropriate. No woman scientist that I knew would have caused such discord; acting unprofessionally would ruin our careers. We wanted to go to sea, to work with our hands, to be treated as equals. It was remarkable that we were even participating on this expedition. Still, the restrictions were absurd. We were instructed to travel in pairs. If we couldn't find a roommate, we couldn't sail on the ship. During *Cocotow*, Karen and I were assigned sleeping quarters on the upper decks. We were not allowed below the main laboratory level. The reasons for the separation were never made clear. Either we were not safe surrounded by men—or we would seduce them. Were we the victims or aggressors or both?

Tanya Atwater, the pioneering Scripps woman oceanographer, later stated that it was years before anyone realized that women were far safer bunked down among the men than sequestered and berthed elsewhere alone.

Karen and I were assigned to the O2 level in an isolated forward cabin that rolled like crazy. I yearned for a lower cabin that was more stable, but there was no chance. We had one breakthrough, however. Kathy Poole, the computer technician, managed to grab a cabin below decks because she hadn't requested an assigned berth beforehand. She just selected an available cabin—first come, first served. She made history for American women at sea. I believe she may have been the first woman allowed to bunk below decks. I envied her because she berthed down where the fun was, while Karen and I were relegated to berths that rode the seas like bucking broncos.

Besides the unusual crew that John Mudie had assembled, the ship had its share of colorful sailors. I have sailed with all kinds of seamen and women, but *Expedition Cocotow* introduced me to one of the most unusual: Bamboo. Bamboo was nearly sixty years old and scruffy as an unmade bed. His onboard tasks included swabbing, chipping, and painting the decks from morning to night. He surprised me one day as I sunned myself on the ship's bow. I wondered at first if he might accost me. Instead, he spoke in a soft voice and asked me about existentialism, whether or not one acquired this philosophy after spending years on the unforgiving blue ocean. Bamboo read volumes of poetry and literary criticism. At every port, his issue of *New York Review of Books* arrived. He stuck the oversized magazine under his dirty arm and stole away to his dark cabin. Bamboo talked for hours. While swabbing and chipping the decks, he drafted in his mind chapters for a book he called "Nine Knots to Nowhere," his autobiography. He fashioned the chapters knot after knot, month after month, year after year. Bamboo was

the only person with whom I felt I could talk. My only other out-
let was my logbook. When I read back through its pages, I am
amazed at how much I learned that first month at sea.

August 26, 1974
Bamboo summed up our expedition by advising me that "there is
no healthy man, only men who are biding their time until they are
ill. So it goes with this ship. There is no fixed or ready ship, only
those that are resting until they break." This ship is a hunk of
miraculous automation that refuses to remain automated. Finally,
we got the depth recorders working again. They had run perfectly
before somebody decided to overhaul them after their initial trial
run on the last cruise.

August 27
We're heading straight for Hurricane Maggie, winds up to 120 mph.
Someone asked, "What good is a ship if it can't get through a
storm?" But I thought, what good is a ship if it can't come *out* of
a storm? Bill Normark and I discussed our plans last night, since it
looked as if two storms would cut off all possible methods of
approach to 21° N. We'll have to hole up near Cedros Island. We've
decided to use the Deep-Tow. Our estimated time of arrival to the
island is 2300 local time, another unexpected night out of bed.

   The entire 12–4 watch in the afternoon was a veritable zoo. People ran in and out with nothing really to do. Things are quieter now.
I anchored myself on deck, life preserver intact and temperature
probe in hand, trying to keep from smashing the very instrument I'm
trying to protect. All seems to be running well with the Deep-Tow.
We conducted two camera runs in Bill Normark's Ranger Slide
area—a huge underwater landslide on the continental shelf of Mexico, and the temperature signal came through beautifully.

I went to bed for two hours and returned to the lab just in time for the pictures of the seafloor to be televised. The gray and white still images slowly emerged on the screen, and the shape of a muddy seafloor arose. Because we have no bottom-anchored transponders, we are trying to navigate our way by watching the radar images of the Mexican coastline. However, when satellite positions come through, the computer plots tell us that our ship is on *land*. I can see that mapping the true shape of the earth is still in its infancy. Still, we continue our survey until the hurricane blows over and we are free to head south again.

August 28, 29, 30

The days are melting into a collection of weird hours of work, watch, and sleep. The temperature system worked well, and we finally pulled the fish out of the water, with the go-ahead from the captain to proceed south to our new location over the East Pacific Rise. Karen's net was in tatters. She compared it to a flag from the American Revolution, a few stars and shreds of blue. Later, we patched it together piecemeal from canvas we located on board. With that done, I sat down with my heat flow instrument and soldered in the correct resistors.

The day was fantastically clear. After dinner, word spread that porpoises were playing around the bow, so I climbed down to the bow dome. The porpoises streamed and zipped. They flipped and turned, swiveled up, shot straight for the surface, leapt in unison, and plummeted back down to the smoothly cascading cleft of water cut by the ship's prow. It was the best display of synchronized swimming I've ever seen.

Afterward, I went to the fo'c'sle [forecastle] to sit and watch the stars. The moon spread a trail of the most enticing gold, straight out in front of us. The wind blew lightly, whispering up sprinkles of water, tiny glittery spheres of phosphorescence. It was heaven, standing there at sea under the sky, feeling the subtle hum of the ship beneath my feet.

Today we woke up to a glassy swell, the winds blowing northward. A blast from Hurricane Maggie flattened the crests of the southward swell. A huge mahimahi arced out of the water. We're well within sight of Baja, so we are fortunate to see a constant array of jutting cliffs against the fluffy mounds of clouds piled high on top of one another. Large schools of flying fish jump out of the water. I try to see what's chasing them.

For the first time in my life, I saw the green flash, that rarely glimpsed phenomenon that happens just as the sun sets over the open ocean. It happened while we were sitting on the fantail singing and playing guitars while the porpoises played in the gilded waves. The sun sank slowly, lighting up the crests of the smooth glossy swells before it dropped below the horizon, leaving in its wake a painted fan of clouds—first a ray of pink, easing to a vibrant purple. The flash was sudden, as the very top arc of the sun disappeared. It was a radiant, lime-green burst. Cheering went up over the entire ship. Almost everyone managed to see it together, at the right moment.

Afterward, I sat up on the flying bridge in Steve Miller's hammock and watched the nearly full moon drift lazily across the cloud-scudded sky. Lightning flashed on the horizon, and a wet warm wind blew from the distant storm. It's probably the tail end of Maggie. During the night, the outer fringes of the hurricane overtook us. Although miles away, Hurricane Maggie's swells cast the ship about in the sea as if it were nothing more than a toy. I could barely keep myself in my bunk. I kept bracing my knee against the ceiling to keep from being ejected over the safety bar. However, seeing the crew walk into the lab or the mess hall at a 45-degree pitch was truly astonishing. At dinner, plates careened away from us faster than we could catch them. Everyone got the food from their neighbor on the left. I was less lucky. I sat next to the bulkhead, and no one's plate came my way.

August 31, September 1

The pitching is now so violent that I cannot sleep. It took all my strength to keep my body in the bunk. I crammed my knees against the ceiling to keep from being catapulted through the porthole. I considered tying myself up with the bungee cord, sort of stringing myself up on the rack. Martin Benson suggested that I borrow more blankets and stuff them in all the open areas around me. Despite the storm, the following day we put the fish in the water again with Karen's net attached to it. I came on watch bleary-eyed and exhausted but managed to navigate ferociously. The art of navigation is similar to war games played by an aging sea captain: "Red ship was blasted off the records! Zing, there goes blue!" I was so tired after watch that I crept up to the steel beach in the morning sun to snooze in Steve's hammock, which miraculously removed practically all of the ship's rocking and rolling. That pure blue sky accompanied by a gentle, warm breeze and the heat of the sun on my forehead soothed my battered bones and weary mind.

September 4

Is today the fourth of September? At least three days have merged into one.

Whenever we are running experiments using the Deep-Tow, it seems as if I am in a constant state of mental elation and physical exhaustion. I am now the regular transponder navigator for the 4 to 8 watch. At one-minute fixes, it takes every ounce of concentration and self-control, especially as the 4:00 A.M. hour approaches. Streaks of color cover my hands from all the unguarded Pentel pens I grab during the four-hour stint of furious, nonstop scrawling. I take this to be the true sign of a navigator.

Meanwhile, the first photos and temperature data ever taken over the East Pacific Rise at 21° North are significant. Our excitement mounted as the temperature kept rising as we sank the instruments further into a fault valley near the spreading center of the rise, down into the rift on the seafloor where the crustal plates separate. The photographs were astonishing.

I stayed up for 24 hours straight to see all of the returning images. They were fantastic, as thrilling as the first photographs of the Moon. Fractures careened across the ocean floor and looked like cracks diving into an inferno. The excitement grew palpable the closer we got to the bottom, and then, in the next second, the photos stopped. No more images were transmitted. We thought we had lost the strobe. The thermometer was still reading, however, and electrical impulses indicated that all devices were still intact, including the strobe. We were only five minutes from the center of the volcanic rift valley.

Still, I was elated. We were collecting incredible data. At 3:00 A.M., Dana came in and shook me awake. I was really gone, with only three hours of sleep in more than two days. We spent the entire watch untangling the various scrawled transponder traces. The fish 5 [Deep-Tow 5] was sent down to take underwater pictures. Without a concern for the strobe, Jack Donovan managed to fly the fish easily. And he grew more daring. The pictures that were transmitted showed the strobe dangling down in narrow caverns; it looked like a submarine spelunker.

September 5

We ended Deep-Tow operations at 0430 Z in the morning. During the night, we practically had a party in the main lab. Almost twenty people were up at midnight, gawking at the pictures of the volcano on the seafloor below. Using the underwater camera, Tony Boege-man snapped pictures right and left. And the photos were incredible: fissures, pillow lava, talus slopes, strange species of fish and sponges, until the fish 5 hit the bottom. At sunrise, I stopped navigating. The fish 5 was really torn up. We had lost one camera, the up-down hydrophone was wobbling, and the front sled runners were bent 45 degrees downward. The fish must have been caught underneath an overhanging ledge and then yanked upward. Still, my data showed that the rise axis is at least warm. In general, temperatures seem to be about a fraction of a degree ($0.05°C$) warmer than temperatures outside of the axis at the same depth. I know many will say that this signal is so slight as to be nothing more than noise. However, the oscillating quartz thermometer has a far greater resolution than the signal I detected. No one had searched systematically for hot water before, here at 21° N, where the axis is active. I wonder what more we would have measured if the Deep-Tow had not crashed, and we had gone all the way to the dead center of the spreading zone. Someday, these data will be very valuable. I know it.

September 6

Cliffs covered with vegetation and islands topped with blue-white puffy clouds dot the coastline of sultry Manzanillo, the banana and malaria capital of Mexico. The water has turned green, and our world has turned heavy, like the dark stillness just before the storm.

Bill Normark and Tony Boegeman will leave the ship. Our origi-nal chief scientist, John Mudie, has arrived. From the flying bridge, we noticed him pacing the dock. Off on the starboard, I saw the Scripps ship *Agassiz* in port with fellow graduate students Jan Hilson and Ted Dengler on board. I didn't expect to see them for another three months. Jan met us and waited on the pier while the new chief decided to give us shore leave. Mudie let us go, and we headed out for a night on the town. First, Jan and Ted gave me a tour of their ship. I couldn't believe the conditions in which they were living. Compared to the *Agassiz*, the *Melville* is a palace. Their corridors are dark, dank, and littered with stray mattresses, scientific equipment, clothes, and sundry other paraphernalia. The heat inside was stifling. The bunks were ancient, stationed beneath round portholes, encased in wooden supports.

We scrambled out of the *Agassiz*, hopped a bus to Santiago along the northern coast, and rumbled past flamingoes and spoonbills feeding in backwater lagoons filled with mangroves in the low tidal flats. Coconut palms, papayas, and mangos dangled over the road. We turned down a small dirt road that coursed through a copra plantation. Thatched houses opened onto the street; chickens, pigs, dogs, and children scratched everywhere. We ended up finally by the sea in a small, protected cove surrounded by palms and dotted with small rock piles. We slipped into the water—it was silky, warm salty freedom. For two weeks, I had dreamed of this. While at sea, we sailed over the most beautiful clear, blue ocean in the world, yet we could not swim in it. Now it all seemed worthwhile. We swam until the hot sun sank below the horizon and the evening star burst like a signal flare over the peninsula.

My logbook entry ended that day, but what followed remains clearly etched in my mind. I was in a pensive state when I returned to the *Melville* much later that night. I knew I was about to face one of the toughest experiences of my life. John Mudie did not seem to be the same person I had known at Scripps. If he was experiencing stress, our expedition surely would exacerbate it. I was the only person remaining on board who knew the original science plan. It would be my responsibility to make sure the plan succeeded. As a first-year graduate student, I couldn't imagine that Mudie would listen to my opinion, but I needed data for my thesis. Above all, I wanted to use the onboard temperature system to search for the hot springs on the East Pacific Rise. I knew that everyone on the ship would be watching me, and I decided that if I performed well, my future at Scripps would be secure. If I sat by and watched as the entire expedition fell apart, I would never forgive myself, and I was convinced that Scripps would never forgive me either.

I didn't sleep that night and stumbled out on the deck in the early morning light. It was my twenty-third birthday. The pelicans were lined up along the cliffs of Manzanillo, and like a squadron, they beat their heavy wings in unison like a sine wave over the pearly white-foamed sea. A few palm fronds drifted by the hull of the *Melville,* and the Mexican workers appeared on the dock, wraithlike, one by one.

8

DANTE'S HOLE

It is characteristic of wisdom not to do desperate things.
—Henry David Thoreau, *Walden*

L ittle did I know that when we headed onto the next leg of the
expedition, *Cocotow 1B*, my last few hours of freedom would
end in a torrent of trouble. I shouldered my responsibilities as I
reviewed our survey plans with John Mudie. It would take a few
days at sea before we reached the Siqueiros Fracture Zone
launch point, and we would need that time to set up the equipment. We steamed out of Manzanillo by 10:00 A.M., the heat of
the tropical sun turning the steel deck into a sizzling hot plate.
By the end of the first day, our tense chief scientist began pelting
us with commands:

"Don't write on that blackboard. It's mine!"

"Don't call up the bridge without my permission!"

"You will stand watch my way!"

"I don't have time to look at the bathymetric maps! I don't need
to see the data!"

"Nobody touch the squawk box! That is my communication
box!"

"There will be no fraternizing with the ship's crew!"

My logbook entries revealed the following:

September 7

After sundown, Mudie came into the lab and began to order people around. Dana, one of our "encounter session" guests, was sent to the computer to assist with the design of his onboard newspaper. I was ordered to stand two consecutive watches in the main lab, the 0–4 and 4–8, and Ron Moe and Steve Miller, our only two computer and navigation specialists, were ordered to stay out of the computer room. As we head more than a thousand miles out to sea, I wonder if we will fall into a nightmare world.

September 9

We are spending $6,000 a day as we head for 8° N, thousands of miles away from San Diego. The chief scientist doesn't seem to remember on which ocean we are sailing. Today he sent a Telex to Scripps (which is copied to all other oceanographic institutions) stating that our work in the *Atlantic* is going well. I wonder, how will we get this science program working? Will I have to take the responsibility?

September 10

Our scientific party has been meeting to plan a course of action for the upcoming weeks. We're climbing the walls. The chief seems sure that a mutiny is brewing among the scientific party. The repercussions could be disastrous for us all. We have opted to inform the captain of everything concerning the scientific work. During our last group meeting, the chief turned his anger on me: "If you don't straighten out, there will be no scientific future for you."

This rattled me. On the one hand, I am concerned for the chief scientist. However, he has blocked every effort to get the Deep-Tow program underway. Later that day we arrived at the intersection between the East Pacific Rise and the Siqueiros Fracture Zone.

Our goal is to determine how a spreading center (one type of plate boundary) and a fracture zone (another type of plate boundary, similar to the San Andreas fault) interact. This will be the first time a Pacific Ocean fracture zone will be studied in detail.

First, we have to construct a map of the area and locate the highs and lows on the ocean floor topography to decide where to place the transponders. The transponder positions are crucially important. A transponder generally is anchored so that it floats 50 to 100 meters above the seafloor. However, if the transponder is lowered on one side of a huge underwater mountain, and we Deep-Tow on the other side of that mountain, we will not be able to navigate since signals between the transponder and the underwater sensing equipment (the fish) will be blocked.

I talked it over with Vince and decided that we need at least five transponders to cover the target territory. During the 4–8 watch, I completed the surface ship profiling and set up five transponder locations inside the deep basin at the intersection of the spreading center and the fracture zone. The chief didn't show for any of these decisions, but when he finally arrived, I left the lab, despite misgivings, to catch a few hours of much needed sleep. When I awoke and returned to the lab, I discovered that only three transponders had been planted, and only two of the three were working. One of the two functioning transponders was set on the wrong side of an underwater mountain. How did this happen? We are more than ten miles out of range and are faced with an enormous problem.

September 11

Chief scientist Mudie decided not to set any additional transponders inside the basin, and we are receiving only one signal. We need at least three functioning transponders before we can lower the

Deep-Tow. Instead, here we are, floating above the intersection basin, which we have dubbed Dante's Hell Hole, given the deteriorating psychological conditions on board, with two wasted days and no navigation work. The chief claims that we have another ten days, but we know that Scripps has allocated only five.

Today, the chief called me to his cabin to tell me that he could make considerable trouble for Peter Lonsdale, Kim Klitgord at Scripps, and Roger Larson, a scientist from Lamont-Doherty Geological Observatory who was slated to run the next cruise on our ship, if they didn't agree to give him extra days at sea. Finally, he wanted me to sign a message to be sent to Scripps requesting an additional week. I said no. I left Mudie's cabin as fast as I could.

September 12

During the night, the chief came into the lab and threw me off watch. I returned to my cabin wondering how I would ever make it through this mess. At dawn, I went back to the lab hoping that the chief had launched at least an additional transponder during the night. Unfortunately, he had not. Instead, during the night watch, he had told the engineer that he was tired and couldn't bother with launching transponders, while several of us who could have helped with this operation were confined to our cabins. At this point, the other members of the staff and I took action. We called the chief's cabin and asked him to launch another transponder because we were in a good location. He answered, "Later in the afternoon, or night." We reminded him that we only had two days left to Deep-Tow. There was no response.

In an act of complete frustration, I marched up to his cabin, woke him up, and announced to him that the scientific party intended to drop a third transponder into the basin. If he had any objections, he could find us in the lab. Before he could say anything, I spun around and returned downstairs. Once there, everything blew apart. The squawk box crackled from the bridge: "Don't let her do it! Don't let her throw the transponder in! I know there's a mutiny on board. You'll all suffer. You'll never regain any status again."

Phil Stotts called back, "Chief, you get down here and live up to your responsibilities, or we'll take drastic action!"

The chief mumbled something inchoate and instead directed Phil about how to address the bridge properly. "Phil, you say 'Bridge, main lab, this is Phil.'"

The ship's captain heard it all. When we switched our squawk box to the bridge and listened in on their conversations, we could hear the captain talking quietly to the chief. Even though, we had confided previously to our captain about the problems in the scientific party, I don't think the captain fully realized how serious our situation was—until now.

After we conferred among ourselves, we sent Vince Pavlicek and Daryl Kinney, two of the Deep-Tow engineers, to the bridge to speak with the captain. Shortly thereafter, Mudie stormed into the lab and summoned all of us to the radio room two decks above. The captain sat on one side, and Mudie sat on the other. The rest of us filled any available space. The captain had arranged a radio patch to Vic Anderson, the acting head of Scripps's Marine Physical Laboratory (MPL), and all the San Diego–based MPL staff. The patch went through, and we all heard Anderson's voice crackle, asking the chief what had happened. Mudie rambled about the newspaper he was writing, the clever editorials, the labeling systems he had implemented for the laboratory instruments, and the trip to the survey site. He said nothing about the Deep-Tow operations. Then, he stopped, offered his resignation, and asked Anderson to appoint a replacement.

Anderson was quick to respond: "Put Benson in charge of engineering and Crane in charge of the science." The silence in the room was deafening for a moment, and then we were asked to leave. All further communication would be conducted in Morse code. The captain took over communications, and we retreated to the ship's library. There, the chief unleashed his fury as he attacked us one by one. Vince interrupted to defend himself, but the rest of the session devolved into a series of murky accusations and denials. Finally, the storm dissipated, and I was left with nothing but a sick feeling in the pit of my stomach.

September 13

During the night, Mudie became violent. He barged into Steve's cabin, spinning and sputtering. He needs to be restrained. We are frightened of what he might do.

The captain now has an active interest in all operations and authorized me to do anything necessary to get the job done. I see to it that we have all transponders in place. We finalized their positions and merged them with satellite fixes. We lowered the Deep-Tow in the basin and started work at last. I haven't seen a star or a sunset, and I haven't slept more than three hours at a stretch, but I have to keep this pace until we finish the survey.

September 14

At last, we have a camera in place along the southernmost reaches of the East Pacific Rise. On the video monitor, I spot a strange form of basalt (volcanic rock), not the bulbous pillow flows to which we are accustomed seeing all along the mid-ocean ridge, but a thin, rippled flow, or ropy, coiled flows found on Hawaii (pahoehoe lavas). This is strange. Pahoehoe is not supposed to occur on the ocean floor. This is a new discovery.

All is now proceeding: The thermometer is functioning properly, Karen's net is operating as commanded, and the photographs are spectacular. Suddenly, at 3 A.M., the chief barged into the lab and screamed at me to leave. Jack Donovan, who was flying the fish, was furious but remained steady for us both: "Kathy, we're coming downhill. Look at the thermometer—now!" We managed to ignore the chief. In spite of the intrusion, our morale is buoyed. Members of the team are carrying twice their expected load, and we are accomplishing an amazing amount of work.

September 15

Today at 11 A.M., Mudie disappeared, and we were happy to be left alone, though not for long. The main lab squawk box sounded, "Main lab, this is the chief. Can anyone tell me where I am?" We wondered, how could Mudie not know where he was? What has caused someone as brilliant as our chief scientist to become so disturbed? Is the oceanographic lifestyle or the stress of the 1970s breaking him apart? Why hadn't anyone back at Scripps noticed that he required help before this cruise?

September 17

After completing a lowering of the Deep-Tow, we needed to take rock and sediment samples from the center of the intersection basin and from the fracture zone ridge just south of the East Pacific Rise. We started at night.

The tropical night spread calmly, like dark syrup over the ink sea. The constellation Vega and the Southern Cross shone brightly above the velvety ocean. Pierced by the ship's spotlights, the transparent water writhed with fish. Needlefish and crabs swarmed around the light. Flying fish whacked against the ship's hull, and once or twice one would whiz through the air, crashing into someone's chest or head. All this happened as we triggered the cores and lowered them into the sea, startling ten-foot-long sharks that were drawn to the ship's nightlights. One after the other, the cores were lowered and raised, their sediments extracted and stored in a freezer. By dawn, we had three good core samples: one chocolate carbonate ooze, one green clay, and one full of manganese chips. When the sun rose, burning off the pink haze, I turned in.

A few hours later, I awoke to the hot steam of the tropics and the last scientific operation we would conduct: a rock dredge for Jim Natland and his colleague Rodey Batiza, a fellow student at Scripps who was studying the origins of seamounts. While the dredge was down, snaking on and off the ocean bottom, its chain link basket snagging anything loose in its path, one of the crew fished off the ship's fantail. He caught a shark by accident, and three people helped haul it onto the deck where it writhed and slapped everything in its proximity. Word traveled around the ship, and in the excitement, Dana, the student volunteer, ran out of the lab to get a look, tripped over a cable, and snapped two of her fingers in the grating on the deck. This roused our chief scientist, who leapt out of his cabin onto the deck shouting, "Turn on all the TVs! Monitor all TVs!"

The chief ran smack into Dana, which threw her into hysterics. We called the captain, who moved Dana promptly to the sick bay. Because we were 1,000 miles from land and there was no medical doctor on board, Bill Siapno, a visitor from Deep Sea Ventures and the only person on board with any real medical experience, radioed a nearby navy ship for information about sedatives and for instructions to set the compound fracture in Dana's hand. Dana screamed pitifully from sick bay, pleading for Vince, or Daryl, or me, but the chief wouldn't let us near her: "You are all cancers. Get out of the corridor!" We shoved past him to reach Dana, to calm her and to reassure her that all that mattered was her health. This young volunteer was also terrified that she had damaged the expedition.

Meanwhile, the dredge returned from the deep, filled with rocks from the seafloor. As we descended to the deck to unload the samples, the chief shouted again, "Get off the fantail!" We turned and asked, "Why?" to which he replied, "Because I'm the chief, and I said so!"

That was it. If I had been any closer, I would have belted him. The entire crew gathered around, twenty scientists and twenty-five crew members, all smelling blood. "Crane!" he shouted, "You count every single rock!" When I told him to count them himself, the whole crew cheered. The chief then lit into the crew. "Hey you! You took some rocks! Show me your rocks!" True to sailor tradition, they laughed and started to unzip their trousers.

Then, from the bridge, the bullhorn blasted: "That will be enough. All hands, back to your stations. Clear the decks!" Our captain put an end to the momentary insanity. Quiet descended on the fantail where a handful of scientists and I sorted through the precious rock samples throughout the night. We hoped that Nat and Rodey would appreciate what it took to get them.

Just before dawn the captain came to me and apologized for all the misery we had experienced on the expedition. His sincerity was enough to wipe away the anger and bitterness, but the sadness remained. "The chief is a very sick man," the captain explained, "and I've been deliberating about whether to head into Acapulco. My second consideration is getting your science done, which is why we're here, Kathy. I'll do my best." By that time, all I could say was, "Take us back. We all need help." The captain agreed that we should return to land, adding, "It's been a hell of a trip. Take her to the barn."

As the captain returned to the bridge, a meteor shower lit up the sky over Dante's Hell Hole. But we were going home. Bill tended to Dana's broken hand as we steamed toward the port of Acapulco. We sat on deck, watching the sky alternate between darkness and daylight, staring into the blue glass sea, trying to pull ourselves together. By September 20, the sickly sweet smell of land wafted over the greening waters. Soon, a tropical bird flew overhead, and then we spotted a soaring black frigate bird. Then came the first sight of land.

An ambulance waited for the wounded as we docked. Dana was helped into the vehicle first. John Mudie, the chief, followed. They both would be escorted back to San Diego. Sirens wailed as the ambulance sped the injured parties away from the *Melville*.

I turned and headed for the Mexican mountains, far away.

9

PARATYPHOID

We have gone round and round the hill
And lost the wood among the trees
And learnt long names for every ill,
And served the mad gods, naming still
The Furies, the Eumenides
—G. K. CHESTERTON, *THE WISE MEN*, 1915

Cocotow 1 ended, but my journeys out to sea during 1974 did not. I was scheduled to ride the ship for another month from Acapulco to Panama. We were not scheduled to use the Deep-Tow; instead, we would carry out magnetic surveys and the mind-numbing retrieval of geological sediment cores, one after the other, endlessly. I don't remember too much about this voyage except for the epidemic of paratyphoid that broke out when we were several days out of Acapulco.

A crew member was the first victim, after which it spread like wildfire throughout the entire ship. So many of us were sick that it was hard to find enough people to stand bridge watch. Science watches were reduced to a minimum. I was one of the last to fall, and it happened in the mess hall while I was standing in line for food. The next thing I knew, the captain was waking me up with smelling salts, and I was lying on the deck next to a table. "You sure picked a hell of a time to come to sea this year," was all he said.

I blacked out again and was carried to my bunk. Cindy Lee, a chemist friend from Scripps who had joined the ship for the second leg of the cruise and my cabin mate, claimed that I shook so hard during my fever spells that I woke her up and she couldn't fall back to sleep. I remember waking only twice during the entire first week. Once I opened my eyes, and the first mate and the captain were standing over me. The first mate was shaking his head, leafing anxiously through a book, when I heard him mutter, "I don't know what to do."

As part of a long-standing cost-saving measure, Scripps's ships had no doctors on board, and the first mate was responsible for all medical emergencies. On the previous leg of *Cocotow,* he had had to deal with psychiatric illness and a compound fractured hand; on this leg of the expedition, he had to deal with an epidemic that he couldn't identify. The stalwarts who resisted the sickness and those who recovered one by one kept the measurements flowing in the science lab, and we continued on our working course. However, it took me more than two weeks to recover.

Larry Mayer and Steve Huestis stole the last remaining quarter bottle of Lomotil and brought it to me, and after I downed the precious liquid, they performed a ritual get well chorus line dance. But I blacked out again for two days.

When I awoke, the cook was trying to feed me cheese, but my stomach rebelled, and I begged for potato chips instead. I am convinced that these chips saved my life. By the end of the week, I could stand feebly on my feet, and by the next day, miraculously, I was almost back to my former strength. We pulled into the queue of ships waiting to dock in Panama, flying a flag with a black pox on it indicating an unknown epidemic sickness on board. Once in port, we were not allowed to leave the ship until the Panama Canal medical authorities cleared every person and all the food and drink on board the ship.

And so we waited. Day after day, all we could do was play quoits—a game like ringtoss—on the fantail and watch our neighbors on an adjacent minesweeper: French sailors in knee-high socks and Bermuda shorts. The Frenchmen ate dinner by candlelight, underneath a colorful canopy that stretched over their fantail. A waiter poured wine with great panache, while we, only five feet away, stared with longing eyes.

Finally, the medical results arrived: epidemic paratyphoid. Source: cheese from Mexico. We were cleared to leave the ship. Steve, Cindy, and I disembarked on shaky legs and retreated to a jungle estate owned by friends working at the Smithsonian Tropical Research Institute. We didn't do much there except feed cockroaches to their pet piranhas, watch for three-toed sloths in the high banana trees, avoid the battalions of army ants leaving destruction in their wake, and listen to the haunting and mesmerizing sounds of seductive Cuna Indian music as we sat around evening campfires.

10

ISLANDS

"Au revoir, mon general. Au revoir, mes camarades. Vive la France," said French General de Castries during the fall of North Vietnam.

The fall of Dienbienphu on May 7 was the decisive development of 1954 in international affairs and a turning point in post–World War II history. It undermined France's will to continue its eight-year-old war against Communism in Indochina and caused it to accept a cruel armistice at Geneva, Switzerland, on July 21.

There was set in motion a series of events that weakened the alliance of Western nations, bared the defenselessness of Southeast Asia, boosted Red China's prestige sky-high, and, in full view of all the world's fence-sitting neutralists, dealt the democracies their worst blow since the Communist conquest of China.

The Vietminh moved in for the kill.

—RICHARD M. GORDON, EXECUTIVE EDITOR,
THE UNICORN BOOK OF 1954

It was 1975. The battle in Vietnam between the East and the West captivated America. The U.S. military was moving equipment and personnel in and out of Guam to Southeast Asia. I also was invited to this region—to Guam, in fact—but not, as far as I knew, for military reasons.

After my previous misadventures at sea, it was rather surprising that I even wanted to be on board a ship again. My thesis work seemed more and more distant. I had lost my thesis advisor, John Mudie. I had thought that perhaps Dr. Spiess would take me under his wing, but he was in England working for the U.S. Office of Naval Research. I started saving what little money I could, because I had a feeling that I would not last in this profession, if my first year was any indication. I had itchy feet and wanted to walk away from the problems at Scripps, but I didn't want a future in Cold War oceanography, and I couldn't rationalize my way into classified research.

One day during lunch on the lawn by the ocean, I was offered the chance to participate on *Expedition Eurydice* to Micronesia. The offer came from Tom Johnson, one of Wolfgang Berger's graduate students. The mission of the expedition was to sample the Ontong Java Plateau north of New Guinea, not far from the horrors of war in Vietnam. Berger and Johnson were short of watch standers, and

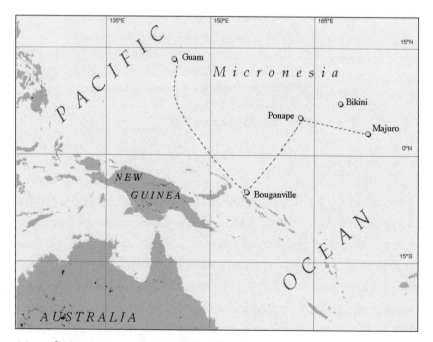

Map of Micronesia and Expedition Eurydice's ship tracks

I decided that it would be interesting to go to Micronesia. The timing was right; I told them, "Yes."

I was hoping that the expedition would provide an emotional break from work on my thesis and that it would rekindle my interest in oceanography. Eight scientists were booked on the ship, and of those, I was the only woman. The trip to Guam was long, since it was one of the old Continental Airlines flights that made stops on several of the Pacific islands before reaching its destination. One of the islands was so small that I couldn't even see it through the window until we touched down. Gooney birds, palms, and runway were all I could make of this spit of land surrounded by the great Pacific. Finally, we reached Guam, the launching pad for military missions into Southeast Asia. I had never been so tired from jet lag. I don't remember much about Guam except that it was filled with American military personnel and had a lovely coastline, some very efficient thieves, a lot of Japanese tourists, and great sushi restaurants.

Scripps's ships often sailed out of Guam and other U.S. Navy bases because the research conducted in the western Pacific was linked, however loosely, with naval missions. The navy would transport Scripps's equipment from San Diego to its bases at no cost. In this case, the navy had donated many sonobuoys as well, and they were stored in Guam.

Once off the plane, we rented a van and took off to go swimming before our forty-day excursion at sea. Unfortunately, we had to park a few miles from the beach, so I left my entire luggage in the van, and when we returned, my luggage was gone. I was now without clothes, except the smelly things on my back, without money, and without my identification cards.

It turned out that it was my job to load all the sonobuoys into a container on the ship, and therefore I had no time to shop before we sailed. Instead, Tom bought my new wardrobe. It wasn't very stylish, but at least I had clothes on my back, and I was grateful that the thieves hadn't stolen my swimming suit, my most important possession at the time.

We pulled out of port on the research vessel *Washington,* leaving the verdant shores of Guam behind, and headed south to New Guinea and Micronesia. Weeks passed before we approached the shores of Bougainville. We had taken a multitude of piston core (cylindrical) and box core (wide square-shaped) samples. In 1975, Scripps's shipboard policy required all expeditions, no matter the mission, to sample water at specific intervals, take XBTs (temperature/depth soundings) every watch, tow for biological samples every day, and run the echo sounder nonstop to gather bathymetric data. To this day, it bothers me that on other ships these samples are not routinely collected while underway. One member of our team sampled the water near Bikini Atoll for traces of radionuclides as a consequence of U.S. nuclear testing there in the 1950s. There were so few of us on board that we had to learn to run all the equipment. However, for a long time, I was not allowed to run the winches or the A-frame, even though this did not require any strength (after all, that is why winches and A-frames were invented). It was not until I made a point of asking that I was permitted to do this job.

One day, we were lowering a core that used doughnut-shaped weights, and so we all lined up to move forty-pound weights from one side of the ship to the core barrel. I was in the line, busy passing weights to my neighbor, when I heard a scream from the main lab, and the next thing I knew, a body had flung itself underneath my arms and was lying on the deck directly under the weight. It was Wolf Berger, our chief scientist. I was so astonished to see him there that it was a miracle that I did not drop the weight right on top of him. Somehow, I had the presence of mind to ask instead whether he needed help. Wolf pried himself off the deck and wiped off his grease-covered shins (he had tripped over a horizontal cable) and apologized. He later told me that he was so shocked to see me in the weight line that he had run out of the main lab to relieve me. After that, he realized that I was handling the weights with ease and that he did not need to protect me from harm any more than he needed to protect anyone else.

The *Washington* rode the seas much better than the *Melville*. It was an old Agor that had been donated to Scripps by the U.S. Navy many years earlier, and its decor was still very institutional, with hatches that you couldn't walk through and ugly green walls. I had a cabin to myself on the main deck, and my schedule was such that I usually slept during the day, when the heat was the most oppressive. There was something very soothing about sleeping with reflections from soft waves drifting around the cabin walls. In the night, which was my long watch, I often monitored all the equipment alone. This required keeping all the bottom profilers in order, taking XBTs every watch, keeping pace with satellite fixes (which came only every few hours or so), and conducting general navigation, which was more of an art than a science. "Dead reckoning," navigation based on course and speed, has pretty much been replaced in the twenty-first century with the "dead-on" GPS navigation provided by the many satellites orbiting the earth. In 1975, the satellites were few and far between, and we kept a log of the times they should be overhead so we could go outside of the lab and watch them glitter across the star-studded sky.

During my watch one night, we came close to Ailinglapalap and Kapingamarangi Atolls. I called up to the bridge to let them know that I was going out onto the fantail to shoot off an XBT. It was about 4:30 in the morning, and the sky was black. The ship was cutting a knife-like wake through the glassy water when I readied my instrument for fire. As I shot it off, the copper wire spooled out into the black sea. We generally had to wait for about ten minutes for the spool to unravel completely, and to make sure that a full-depth record was taken, before we could return to the lab. It was during this wait that I spotted what I thought was a phosphorescent log, about two to three meters in length, probably a coconut palm tree in the water. We were sailing very close to the tree-covered coral atolls. I was thinking how beautiful this log was, drifting so close to our ship, when the copper wire ended and I had to lean over the railing to sever it from the machine. All of a sudden, the log came alive, opened up like an umbrella into many arms, and then shot away into the ocean. It was a huge, glowing squid.

I fell back onto the deck in astonishment. Stunned, I entered the main lab and called up to the bridge, "Bridge, main lab; safely back from the fantail."

Schools of tuna raced through the sea and kept time with our ship. During night coring stations, we were usually surrounded by sharks, which circled from one side of the ship to the other. At intervals, we would take deep trawls for midwater fish. Larry would analyze the fish for any radionuclide contamination. The results of our contaminant survey indicated that there was no significant concentration of cesium or strontium in any of the samples, which puzzled us, since we were rather close to Bikini Atoll, the site of American nuclear testing. Many years later, I learned that Japanese scientists had surveyed this region earlier, and they had mapped the great Pacific plume of radionuclides that spread from Micronesia *north* to Alaska. We were measuring the region to the south of the atoll, far from the giant oceanic currents that carried away the evidence of radionuclide contamination.

Only once were we hit by a storm, and it was tropical squall, a real humdinger that seemed to come out of nowhere. We were watching a cowboy movie far up in the bow, when the giant waves hit at the same moment that a bank was blown up in the film. The waves knocked the projector to the ceiling, and it rolled over and over, streaming greatly animated footage of an explosion across every wall of the mess hall. Then it crashed onto the floor, and that was the end of our film.

Shortly thereafter, the cook came down with hepatitis and had to be put off the ship at the closest Micronesian port, Ponape. Our captain, Captain Bonham, my favorite of the Scripps fleet (because he used to take the time to teach us old nautical skills, such as tying knots), was the only person on the ship who had previously been to Ponape, but his excursion had taken place during World War II on a rubber dingy, escaping from his sinking ship that was struck by Japanese torpedoes. He had only the vaguest of recollections about the configuration of the coral reef that surrounded the island.

It was necessary to find an opening in the reef so that we could land. In the 1940s, the entrance for ships was on the eastern side of the island, and we headed to that side and slowly maneuvered the *Washington* up a channel in the reef. I manned the echo sounder to make sure that we would not ground the ship on the coral below. Soon, it was all too evident that the reef had closed itself up over the intervening thirty years, and we would have to find a new route. When I glanced out the starboard porthole, I spotted three canoes paddling in our direction from shore. The people in the canoes were wildly gesturing for us to go back. All engines came to a halt, the canoes pulled up alongside our ship, and we threw down a line to the natives. One came on board to talk with the captain, and he offered to guide us to the northern side of the island where a new channel had been cut recently through the reef. With our native guide on board, we were able to back carefully out of our ill-fated channel and head into the correct entrance to Ponape.

In the harbor, a Japanese fishing boat was covered from stern to bow with lines of shark fins. The fins were destined to become shark fin soup or to be processed into some high-priced Asian aphrodisiac or virility potion. We spent one day on the island, off-loading the cook and gathering coconuts and peppercorns from the local trees. The Japanese fishermen were much in awe of the giant "cannon" that we had strapped onto the port side of our ship. The "cannon" was no more than our piston core, but it surely looked like a weapon. We left Ponape shortly thereafter, minus one cook but loaded with coconuts.

Night coring off Ailinglapalap, Micronesia, 1975. Left to right, Tom Walsh, Robert Wilson, Kathleen Crane, George Wilson, Wolfgang Berger, and Peter Roth. PHOTO: *Tom Johnson, Scripps*

Back at sea, we finished sampling carbonates from the seafloor, deploying the sonobuoys, and measuring the crustal thickness of the Ontong Java Plateau. Then we headed to Majuro, our final destination in the Marshall Islands. Majuro is a beautiful island atoll littered with refuse from World War II. Tanks and landing craft lie rusting on the beaches and in the jungles, but the beaches and lagoons remain otherwise pristine. Never before had I observed such an array of tropical fish nor found such a diverse collection of seashells.

After two idyllic days, I boarded a plane for the U.S. mainland by way of Hawaii. We arrived in Honolulu at about 3 A.M., and I stumbled out of the plane to find something to eat. Hundreds of people, mostly Asians, occupied the seats and lay on the floor, clustered in groups. Some were well dressed, and others looked like peasants. All of this meant nothing to me in my state of exhaustion; my singular goal was to reach the nearest candy machine. I trudged straight ahead, when from my right came the shout: "HALT!"

The barrel of a rifle was whipped out in front of my chest. I was so exhausted that I hardly reacted; I just kept walking to where I believed the vending machines would be. But the guard was insistent. I finally gathered my wits enough to ask what he was doing, and he snarled, "Don't you know what has happened? These are all refugees from Saigon on their way to California. Vietnam has fallen."

11

STORM

It is a quiet summer morning in northern Michigan in the year 2000. My daughter is still asleep in the old quilt-covered porch bed, and the remains of her painting the evening before lie scattered over the floor of our 1922 cottage. The breeze hasn't come up yet, and the lake surface lies still, like glass, as it reflects the white birch trees that lean precariously over the soft pine-needle embankment. Near the shore, my grandfather's rowboat shimmers with reflections of water on its varnished exterior. There is time to sneak in a morning swim before the rest of the world awakes and the daily routine begins. I have come to this cottage by the "sweet" water of the Great Lakes whenever possible to return to the earth, to recharge, and to reflect on my life course. And it was here that I came in 1975 when I needed to gather strength to continue my difficult path in graduate school.

In the spring of 1975, I was halfway through the Scripps Ph.D. program, but I was facing the possibility of completely failing in my efforts to locate deep-sea hot springs. Although the scientists at the Scripps Marine Physical Lab (MPL) were growing interested in my thesis, and I was allowed to pass my oral exams and to redesign the Deep-Tow thermometer before Dr. Spiess left for England, I had encountered many personal difficulties along the way. It was clear that my greatest problem lay with Peter Lonsdale, who had been a friend from the house on Nautilus Street, but who was growing into an adversary, though not by my choice, as he advanced in the world of science. After *Cocotow 1,* when John Mudie had stepped back from daily work at the MPL, Peter, who had just graduated, stepped into the gap. There was no doubt that he was a brilliant scientist, but he was determined to make a name for himself, even if it cost him friendships.

While Dr. Spiess was in England, Peter essentially controlled almost every aspect of the Scripps MPL. During that year, more and more attention was focused on the study of the "missing heat" from the mid-ocean ridge. A scientist from Oregon State University, Jack Corliss, sent a proposal to the National Science Foundation, together with Dick Von Herzen of the Woods Hole Oceanographic Institution, to search for the missing heat at the Galápagos Spreading Center; they were funded. Corliss inferred the presence of hot water from the hydrothermal type of minerals that he had sampled from the mid-ocean ridge. Von Herzen studied classical conductive heat flow. However, no one was actually searching for the hot springs themselves. Bob Detrick, who had been conducting similar research, had resigned the previous year. The exploration for the heat source was a job for a marine geologist, one who could analyze the seafloor topography and water column and use the right tools to home into the precise location of the hot spots. That would be my job, according to my new thesis advisor, Dr. Spiess. With Dr. Spiess's agreement, I asked Dr. Von Herzen to be my thesis co-advisor, as he was a specialist in the field of heat flow—and I needed backup at Scripps while Spiess was in England. During 1975, Von Herzen was on sabbatical leave at Scripps from Woods Hole Oceanographic Institution. This arrangement was perfect for me, as it provided the necessary senior support I needed to carry out my thesis work.

In 1975, Peter had the Deep-Tow under his control, and he knew that it was the one tool that could carry out the initial search and location of hot springs. To anchor his position in this growing field, Lonsdale quickly set up camp with two chemists at Scripps: Harmon Craig and Ray Weiss. Craig and Weiss had been tracing a plume of helium-3 (an indication of hot mantle material) from the East Pacific Rise toward Micronesia. Like Peter, I was a marine geologist. I also had experience with the Deep-Tow, and that was too close to his own expertise. I was Lonsdale's junior, and while Spiess was away, I had little say in the fate of the data I collected for my thesis. Prior to 1975, Peter had never been involved or interested in the study of the mid-ocean ridges. Instead, he was an expert on sediment and mud waves and the role of deep-sea currents in reshaping the seafloor. His focus changed with the anticipation of a new discovery. As it turned out, the discovery of underwater hot springs would be one of the great earth science accomplishments in the late twentieth century.

The Galápagos Spreading Center expeditions were scheduled to take place in 1976. During the summer of 1975, we were scheduled to work on and near the Reykjanes Ridge south of Iceland. I persuaded Dr. Spiess to allow me to try again, to find hotsprings on *Expedition NATOW* (North Atlantic Tow). Our goals would be twofold: sedimentology (mapping the great chasm, the Maury Channel, on the seafloor, which had been created by the overflow of water from massive melting of Icelandic glaciers during volcanic events); and volcanology (exploring the Reykjanes Ridge, a section of the mid-ocean ridge extending south from Iceland).

Peter Lonsdale and Charlie Hollister from Woods Hole Oceanographic Institution were the chief scientists for the expedition, and their primary interest was the exploration and mapping of the

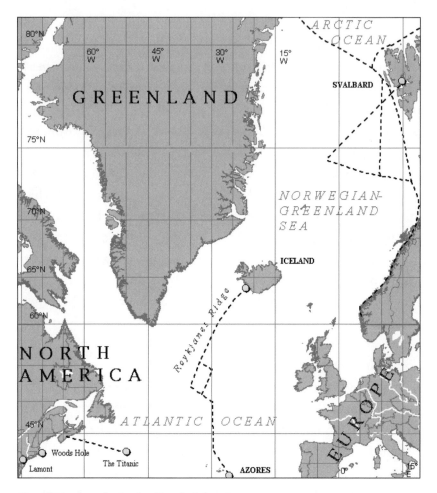

Expedition tracks in the North Atlantic

Maury Channel. The head of the Reykjanes Ridge search was
Tanya Atwater from MIT—a former student of John Mudie.

Difficulties arose long before we left port. In 1974, with Dr.
Spiess's approval, I had asked Vince Pavlicek to design a way to
lower the Deep-Tow's quartz thermometer below the fish so that
our chances of finding the warm water would be greater. This was

the configuration we had used during the search for hot water over the East Pacific Rise. However, as soon as Peter became involved in the hydrothermal search, he prohibited the Deep-Tow engineers from lowering my equipment from the fish. There was no one to arbitrate this conflict because Dr. Spiess was away, so we ended up going to the Reykjanes Ridge during the summer of 1975 with the Deep-Tow thermometer strapped high up on the cable. Peter's work was still focused on the deep-sea sediment canyon systems, and the working relationship between the sediment team—Lonsdale and Hollister—and the ridge group led by Atwater was tense, to say the least.

Prior to leaving Scripps, I had asked Scripps physical oceanographer Joe Reid if I might borrow a current meter that I could deploy on the flanks of the rise axis. This would be dropped onto the seafloor east of the Reykjanes Ridge and would gather information about the direction and magnitude of the bottom currents in the area. One problem in searching for hot springs on the seafloor is that the temperature signal is often very, very small (on the order of a few hundredths of a degree). Two currents flowing in opposite directions with substantially different temperatures ran along the flanks of the Reykjanes Ridge. Some of the water spilled over the ridge through low points from west to east and therefore "contaminated" the temperature signals when water masses of different types mixed together. Trying to find a hot spring in all this activity was really like finding a needle in a haystack. The signal was often much smaller than the surrounding noise. Data from a current meter and a CTD (which measures conductivity, temperature, and the depth in the water column) were necessary to document the background temperature, salinity, and velocity signals. Joe Reid was also on my thesis committee, although he himself did not believe in the existence of hot springs on the seafloor. He had already told me, "You can alter the oceanography of the seafloor with your so-called hot springs, but don't touch my layer-cake stratified ocean-ography of the surface ocean." He loaned me the current meter anyway. I shall be always grateful.

Tanya Atwater had invited a Soviet marine geophysicist to par-ticipate on our expedition. It was a surprise to Dr. Spiess, and when he learned that a Soviet scientist was on board, we received a secret message from London that said, "Don't tell him anything about the Deep-Tow."

A Soviet fishing trawler followed us every day of the expedition, and only then did I realize that the region we were studying was politically sensitive. The Reykjanes Ridge, just south of Iceland, was the focus of many U.S.-U.S.S.R. submarine cat-and-mouse exercises. We were funded to gather ocean-bottom data that could help submarines hide behind cliffs or in deep troughs bordered by faults. This is how the U.S. Office of Naval Research justified much of our "pure" research. It was not something that I liked to think about, and it would be many years before I would directly address the issue of Cold War–driven ocean research.

Overall, our team comprised an impressive array of Deep-Tow scientists, past and present. Tanya Atwater, John Shih, Sandy Shor, and Charlie Hollister were from MIT and Woods Hole. Peter Lons-dale, Larry Mayer, Chuck Alexander, Marcia McNutt, Kathy Poole, Steve Miller, Karen Wishner, and I were from Scripps, and all the work was to be performed on the Woods Hole research vessel *Knorr*. Tanya, John, and Marcia were studying the magnetic pat-terns associated with the plate boundary evolution of the Reyk-janes Ridge. I was searching for hydrothermal vents. Peter, Sandy, and Larry were focusing their work on the Maury Channel and other seafloor sedimentary transport pathways of the deep North Atlantic Ocean.

This was the first time that I had encountered a crew that was so anti-female *and* anti-scientist. Before leaving port, Captain Hiller lectured all the women about how we should behave on board. Unfortunately, this was getting to be a routine performance from expedition to expedition, and the women were growing tired of this admonishment. I could not be more robotic than I already was. Furthermore, each member of the Woods Hole crew had his own position and would not relinquish it for anything. At Scripps, we normally carried out all the launch procedures ourselves, separate from the crew. The students would line the deck, holding the cables taut until the termination switch clicked in to the Deep-Tow and it was lifted safely from the deck. The chief scientist usually directed the launch procedures, choreographed the winch operators, and made sure that all the parts worked smoothly. The *Knorr's* crew, however, did not like this arrangement. The ship's fantail was their domain, and the scientists were expected to respect that by staying out of the way. But this would not hold true for this expedition. We managed as a team to break through the barriers that existed for scientists and for women.

The cruise left from the Azores, islands located in the central Atlantic and controlled by Portugal. We were docked next to a huge Soviet ship, which was carrying hundreds of students from Cuba to Leningrad. The port was a bizarre mix of bananas, and rum, vodka, and whiskey, and Russian, Spanish, English, and Portuguese. The Azores are a main stopping-off point for scientific expeditions in the North Atlantic, and their lush environment would contrast greatly with the violent North Atlantic storms we were heading into in the coming month.

Rolling along the "highway to hell," in the North Atlantic, south of Iceland, on board the Knorr.

Working in the seas around Iceland is my vision of a cold hell. Storms come out of nowhere, and they build up momentous force in a manner of minutes. The storms notwithstanding, we spent several weeks chasing the Feni Drift and the Maury Channel east of the Reykjanes Ridge, and finally, the time came for our transects up the ridge axis to Reykjavík, Iceland. The days were very long and followed one after the other in unending sameness. To relieve the monotony, we played Scrabble and Ping-Pong and tried to read amidst the huge seas. Late one night, finally on our way to begin our axis survey, Marcia McNutt awakened me. "Kathy, the chief scientists have thrown over your current meter into their research area. You'd better come on deck right now."

What could I do? There was not one chief scientist on board who was backing the work on the volcanic rise axis. Again, it looked as if my entire survey plans would be scrapped, because they would not be retrieving that current meter for another month. Without the current meter, I would not be able to decipher between the signal of a hydrothermal vent and the background noise. Instead of fighting, I gave in, and we continued with the other research projects. I still had the Deep-Tow thermometer, and I maintained hope that I could take measurements at some point in the expedition. However, the North Atlantic storms would dash those hopes.

Shortly before reaching our destination at the Reykjanes Ridge axis, we encountered engine problems and were unable to proceed at speeds less than 4 knots. We scrambled, recast our plans, and decided to take the Deep-Tow to the axis, zigzagging up its length. Toward the end of the axis, a huge storm erupted, and our ability to navigate nearly collapsed. We couldn't point our bow into the waves and wind, and we were forced into the impossible situation of taking enormous seas broadside against the ship. This is the most dangerous situation to be in at sea. Everything on the ship started to fly around. Objects, even those tied down, upended and broke away from their constraints. The Ping-Pong table let go, books flew off the library shelves, and no scientist could stay in a bunk. We all huddled together, lying next to one another on the floor of the main lab, hoping that somehow the crew would pull us through.

A ship's crew is often called upon to perform miracles. On this leg they were able to "gin up" a rudimentary replacement of some needed engine replacement parts, and we limped slowly toward Iceland. During the next leg of *NATOW,* the patchwork repair fell apart again, and the crew had to use a huge supply of canvas on board to construct a giant sail, which they strung up partly on the bow. The once proud Woods Hole research vessel *Knorr* came sailing into port, its engines nearly destroyed, its scientists and crew exhausted and devastated.

The *NATOW* expedition did not generate any advances toward unraveling the great undersea hot spring mystery. In fact, we were so unsuccessful from a combination of negative human and meteorological conditions that Charlie Hollister, the Woods Hole chief scientist, told me late in the summer of 1975 in Iceland, "Kathy, you'd better find a new thesis. You'll never find hot springs on the bottom of the ocean." I wish I had that statement on tape. It would prove to me that I was wise not to give up, for even senior scientists are not right all the time.

By late April 1976, I was ready yet again to search for hot springs, only this time on a Scripps ship heading for the Galápagos Spreading Center. I would spend two months with a scientific party assembled from the Scripps Deep-Tow Group, teams from Woods Hole and Oregon State University, and most important, one of my thesis advisors. Dr. Spiess was back in town.

12

THE BEST OF TIMES,
THE WORST OF TIMES

And he lay looking at the map for
Five years more before he
Saw that it showed the way to eternity
—FLANN O'BRIEN

Toward the end of the 1970s, Cold War–driven oceanography was changing dramatically. There was a revived competition to make new oceanographic discoveries, which allowed new people with new ideas to rise rapidly to the forefront of the science. In some ways, I benefited from these changes. However, Western oceanography continued to base its existence on finding ways to beat the Eastern bloc. The military still partially controlled the focus of our work, and competition was fierce. It was so intense that some of the men in our field grew frustrated with the likelihood that they would have to cede some of their success to the new and growing group of talented female oceanographers. The Scripps Deep-Tow was still being used to track sunken submarines, even as our research moved into new directions of undersea exploration. Our group was led by the ultimate "stealth" oceanographer, Dr. Spiess. Of all the people I worked with, I never expected that Dr. Spiess, the hard-core former navy officer, would accept women on his team, yet he became our biggest supporter at the time.

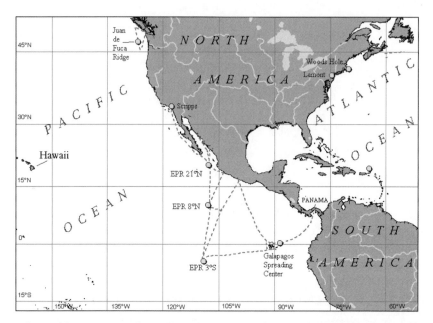

The Galápagos Spreading Center is an area deep in the Pacific Ocean off the west coast of South America. It is the site of the discovery of life at oceanic hydrothermal vents.

By the time *Expedition Pleiades* occurred I was just glad *not* to be going back to the ferocious North Atlantic. Instead, we would head for the East Pacific Rise at 3° S and the Galápagos Spreading Center on the equator. It was here, in 1972, that Bob Detrick and Dave Williams had discovered some tiny temperature anomalies in the middle of the undersea rift valley. Many people, including Bob, were somewhat skeptical about the meaning of these anomalies, but I was optimistic and thought that they meant a great deal.

My goal on the expedition was to locate, take samples of, and to leave navigation markers (transponders) at the sites of hot springs at the seafloor spreading center. This particular site was chosen because it was located exactly 200 nautical miles from the Galápa-

gos Islands and was legally out of the range of Ecuadorian gun-
boats, which were patrolling for illegal tuna fishing in their waters.
At 200 nautical miles, we were in international waters and could
carry out research without the fear of being hauled into port under
the guidance of warships. It was in some ways political serendipity
that led to one of the most interesting scientific discoveries of the
twentieth century.

We had allocated two months for the expedition: one month to
locate the springs with the Deep-Tow, and one month to conduct
sampling using heat flow equipment operated by Ken Green and
Dick Von Herzen and to photograph our findings in color using a
Woods Hole camera sled. I was not only a watch leader but also in
charge of the navigation of the Scripps transponders for the *Expedition Pleiades* second leg. We needed to leave transponders behind
in order to mark the site of the hot springs. This job was of paramount importance, and Spiess had entrusted it to me. My friend
and new Deep-Tow engineer, Bob Truesdale, would work with
me. We left San Diego on the *Melville* in late April under amazingly
calm seas and skies.

During the first seven days of steaming time, I learned new computer code and tried to absorb the information needed to initialize
the Deep-Tow computers so that I would be ready to run the relay
navigation for the second leg of the expedition. The long haul was
arduous, but it was wonderful to see the constellation Pleiades high
up in the sky again. The stars Aldebaran and El Nath would be visible in the sky later on. Because of the tropical weather, I was eager
to take an upper bunk for more privacy. However, there was very little space between the mattress and the ceiling of our cabin. When
I rolled over on my side, my shoulders got stuck, wedging me in,
almost every night. On some of the nights, when tropical rains did

not pound the ocean surface flat, I slept outside under the stars, try-ing to absorb the small pleasures before the hard work began.

By the eighth day at sea, squalls developed over the horizon, and we steered between the tiny storms. In contrast to the caressing tropical warmth, there was a freezing blast of air coming from the main lab where an air conditioning system had been newly installed to keep the computers humming during operation. It felt like the "death of the tropics."

As we passed over "Dante's Hell Hole," Steve Miller, Conrad Young, and Mark Legg took a burnt-out projection bulb and sacri-ficed it to the sea. As soon as it touched water, sparks shot out in all directions. It was a fitting memorial to the tempestuous *Expedition Cocotow.*

By May 11, we had reached the first survey site at 3° S. There was no end to our Deep-Towing problems this time. Ray Weiss's newly built CTD was acting strangely, the transponders were faulty, and the magnetometer winch was broken, forcing us to launch and retrieve streamers and cables hand over hand. At least this was good exercise.

On one occasion, I was pulling in a small hydrophone when a male colleague started to push me out of the way to grab the cable. I am sure that he would not have done this to another man, and since I was tired, I artlessly remarked, "Please move and let me do my job." He stepped aside.

In those days, many women gave in to the male-controlled world and seemed to believe that they were incapable of simple physical tasks. I still wonder why they treated the women scientists as they did. Was it to protect women from harm? Were they merely doing what they had been trained to do by their mothers? Maybe, but it was really hard being the person subjected to a false sort of protection, because I didn't want it, and I was not alone. Oddly enough, the Scripps crews did not subject us to such humiliation; more often, scientist colleagues, especially the older fellows, attempted to keep us out of harm's way. Sometimes these simple acts, well meaning though they may have been, frustrated me. I would retreat to my cabin whenever I could.

During one test run of the Deep-Tow, I accidentally tripped and fell in an attempt to keep out of the way of the chief scientist. It is amazing to me how clumsy I could be when I was afraid of interfering. I found out later that I had fractured my elbow. I so desperately wanted to remain on board that I hid my pain as best I could and bandaged up my arm in a makeshift sling so that it would not move when I was sleeping. The regular hum of the Deep-Tow winch lulled me into pain-free unconsciousness.

We were not so successful with the Deep-Tow on this excursion over the East Pacific Rise, but we did manage to haul in a huge colony of gelatinous animals. These phosphorescent blobs were more that one foot in length. Mahimahi were everywhere, and their rainbow sleek bodies streaked by our ship leaving glittering, iridescent wavelets behind. Some mahimahi wound up caught, fried, and served as dinner.

During this half of the expedition, I knew that I would not be allowed to collect any data for my thesis. As chief scientist, Peter Lonsdale was running the operations for Scripps, and this was his corner of the world. I kept my head down, read up on Deep-Tow navigation systems, and did my job on watch as navigator and watch leader.

Meanwhile, we surveyed continuously. After standing the morning 4–8 watch—the one I love the best—I would meander out to the fantail and just sit on a capstan to drink in the colors. In the tropics, the gentle morning glow could turn rapidly into a raging shaft of light. Equatorial clouds, with their flat bottoms and puffy tops, coursed across the sky in a stately manner. Sometimes they seemed to skitter around. Others hugged the horizon and waited for their turn to rain down onto the sea. A fragment of poetry written by G. K. Chesterton often crossed my mind as I wandered off to my morning sleep:

> The sun on my left
> the moon on my right
> my sister, good morning,
> my brother, good night.

Every time I drifted to sleep, I dreamed of faraway mountains covered with snow and of walking along a trail leading up to them. Dreams at sea are strange. By the fifth week, my dreams just evolved into swirling masses of gray and blue without narrative, people, or adventure, and I was left with only the splash and gurgle of the sea.

On our last day at sea over the East Pacific Rise at 3° S, a beautiful sail showed up on the horizon. For weeks we had not seen another vessel. We had been completely alone, when out of the blue a crimson sail appeared, looking like a dream as it grew closer and closer. The boat so captivated the attention of every single member of the crew and scientific party that we lined up along one side of the ship and stared. As it drew nearer, we took bets about the size of its crew, its origin and destination. Binoculars came out. The crew of two, a man and a woman, were naked. We could see them scrambling to put on their clothes even as they brought their boat in as close to windward as they could, straining against the sheet.

"Where are you from?" we called out.

"Denmark, ninety-six days out."

"Where are you going?"

"The Marquesas."

Soon the crimson sail slipped by our stern and zipped off against the fluttering breeze, and it became a mere wisp on the horizon. What a feeling they inspired, free to sail the high seas wherever the wind would take them. At dinner, everyone seemed quiet. It was as if we had experienced a vision, a religious miracle from out of the middle of nowhere.

By May 26 we reached the Galápagos Islands: Barrington (Santa Fe), Chatham (San Cristóbal), Indefatigable (Santa Cruz), as well as small islets topped with tufted, bushy vegetation. The sky was hazy, but at night the sea glittered with phosphorescence. Once a porpoise leapt up and arched down a full moon beam against the black silhouette of an island in the distance.

The Deep-Tow, with Weiss's water bottles attached, the first sampling of the Galápagos hot springs, during Expedition Pleiades in 1976.
PHOTO: *Scripps*

Before we could look for any hydrothermal vents, we needed to spend interminable days taking core samples for Peter Lonsdale. I manned the capstan. I remembered that John Mudie once pointed out that this was the safest job around, but I told him at the time that I might as well be a button pusher.

The week of May 26 to June 2 was hectic. We launched transponders over the Galápagos Spreading Center, navigated them into position, and began Deep-Towing. A row of bottles designed by Ray Weiss was strapped to the Deep-Tow.

The first clam sightings, Galápagos Spreading Center,
Deep-Tow photo, 1976.

We also attached the CTD to the fish and dangled my quartz thermometer below. On June 2, the Deep-Tow swept over warm water emanating from the fissures in the rift valley below. We recorded a temperature increase of 0.1°C (which was a phenomenal increase given the Deep-Tow's height above bottom), and we trapped bottom water in Ray's bottles. For three days I did not sleep more than two hours. My thesis was being confirmed, but now I worried about the ambitions of the new hot spring converts. We mapped out spring after spring, took bottom photos of stained pillow basalts, glass-covered basalts, and clamshells on the seafloor. I noted that these were located on the seafloor where we had recorded the high temperature spikes. Peter, however, stated that the shells were merely the trash thrown overboard from a clambake probably held on a nearby navy ship. Maybe he was joking, maybe not. It was true that we also photographed plenty of beer cans. In fact, each ship had a different type of beer, so by mapping the litter you could reconstruct the ship tracks of all the vessels in the area. Because of Peter's comments, I called these springs Clambake I and Clambake II.

Despite all the excitement, I tried to stay out of Peter's way. We sent radio reports out to Scripps: "Mission successful!" We left the transponders on the seafloor at the end of this first leg at sea. Peter and Ray (along with most of our team) would disembark in Panama. Bob Truesdale and I would run the transponder navigation on Leg 2 for the incoming Oregon State University and Woods Hole teams headed by Jack Corliss and Dick Von Herzen. Our final duty would be to place two long-life transponders on the seafloor, one on a hot spring and one on the off-axis mud mounds, for use during the first human visit to the hot springs by way of the famous deep-diving submersible *Alvin*, one year later in 1977.

By June 7, we were in port at Panama, and our chief scientists Lonsdale and Weiss left for San Diego. A letter arrived for me from Scripps from my fellow Deep-Tow office mate:

June 7, 1976

Dear Kathy,

Hopefully this note will be hand delivered by Russ & will find you in peak spirits. If not you can just think that the worst is all over & it can only get better from now on. Life has been fairly pleasant here, but I'm sure that will all change with the imminent return of our illustrious leaders. I must have caught something from you because the office has been a real mess since you've been gone. . . .

Reading the radio reports—it sounds like you've found some fan-tastic stuff. We all just hope that you'll be getting your share of the data. It is very exciting . . . almost to the point that I wish I was there, but I must stress the almost.

Love

Larry

By this time, I knew we were on the verge of something really momentous, but I was feeling even more uneasy about the situa-tion. Only a very few people had believed in the existence of hot springs when I started my search, but now a handful of competi-tive scientists with strong personalities had suddenly converted to the idea. These people were senior to me, and they were going back to San Diego. While I would be conducting my assignment on Leg 2 at sea, they would be in the position to announce the results of our remarkable data—without me. And that is exactly what they did.

I was right to worry about whether I would have any data for my thesis. But I knew I had a job to do for Dr. Spiess, and that was to ensure the retrieval of all the Scripps transponders and to deploy the critical navigational tools to bring us back to the site in 1977.

During the second leg of *Expedition Pleiades,* I trained nine peo-ple to operate the relay navigation using the Scripps transponders. I was also in charge of smoothing any navigation errors. Bob Truesdale and I were a very good team. Between his work as engi-neer and mine as software and navigation coordinator, we were able to "reboot" all systems, from the nuts and bolts to computer algorithms, when the relay transponders attached to cameras didn't work. Our teamwork would prove essential since all the var-ious scientists depended upon good navigation to gather samples.

A large squadron of geochemists from Oregon State University and geophysicists from Woods Hole Oceanographic Institution had come on board at Panama to replace the Scripps team. The ship was so crowded that one person had to sleep on a cot for lack of a vacant bunk.

A good friend from Scripps, Russ McDuff, had arrived as well. He was a bright chemist who would examine pore waters from the sediments adjacent to the Galápagos Spreading Center. As classmates, Russ, Larry Mayer, and I looked out for and defended one another regularly. At Scripps, a graduate student needed a good team of defense and support. On Leg 2, Russ and Larry knew of my growing concern about the pending publication of the hot springs data collected on Leg 1. Ever the diplomat, Russ had overheard a discussion between Jack Corliss and Dave Williams. Russ explained to me that Corliss and Williams were eager to publish a scientific paper on the Galápagos hot spring discoveries before Peter Lonsdale and Ray Weiss. To succeed in science, a scientist must publish research results rapidly, and when a new discovery is made, competition among the participating scientists immediately becomes ferocious. However, Peter and Ray were already back at Scripps and had access to the top scientific journals. We were still locked away at sea with no way to communicate. The race was on, and I realized that I would lose out. How could I complete my Ph.D. without those data?

The work at the onset of Leg 2 on the *Melville* seemed ceaseless, day after day, hour after hour. I was so busy that for the first three weeks, the only entry in my logbook was, "Well, finally made it to July!" To keep the navigation running smoothly, I spent hours punching and repunching cards for our giant mainframe computer. On July 4 we launched long-life transponder Green III to Clambake I, and shortly thereafter, we launched the second long-life transponder into the hydrothermal mound region on the flanks of the spreading center. We checked the transponders' communications to make sure they were working, and then we programmed them into a dormant state so that they would answer only to a special frequency-modulated signal that I would send from another research ship, the *Knorr,* sometime the following year.

We still needed to retrieve the four standard transponders before we could close down Leg 2. The scientific party wanted to keep using the transponder net as long as possible to get accurate navigation for their cores and camera runs. If we did not actively retrieve them, the transponders had a built-in fail-safe mechanism that was set to trigger the transponders to surface on July 7 at a specific time. Dr. Spiess had charged me to bring them "all back alive," and it was getting down to the wire. On the night that one of the transponders was to surface, the scientists on watch directed the bridge to take the ship about ten miles from the transponder's location. I was asleep when this happened. Prior to turning in, I asked that the ship remain on the site so that if the transponder surfaced, we would be able to retrieve it easily. When I awoke, I was shocked to discover that we were ten miles away. Sure enough, the transponder had released from the seafloor and was drifting around in the waves. Bob, Russ, and I had no choice but to get to the bridge, where we spent ten furious hours trying to triangulate the ship to the transponder, even as it was flowing away from us by powerfully flowing currents. The only signal we received from the transponder was the "slant range." We could tell that we were within one mile from the instrument, but we didn't know which direction to go. We had to turn the ship in one direction and move steadily to see if we were getting closer or further away. This went on for hours

```
NR  5 WCOB MELVILLE CK 185     0622552 JUL 76

SPIESS, BOEGEMAN, LONSDALE MPL

TWO LONG LIFE TRANSPONDERS SUCCESSFULLY DEPLOYED.  HAVE RECOVERED
4 PLEIADES 1 TRANSPONDERS. R1 R2 B2 B2.  ATTEMPTED TO RECALL B1 FOR

SIX HOURS.  RECEIVED FLAG FOR 1.5  MINUTES BUT NO LIFT OFF.  PASSED
TIME RELEASE, NO LIFT OFF.  B1 TO BE RECALLED JULY 7.  DATA FOUND

ON ALL RECOVERED CURRENT METER RECORDS.  BASELINE ERRORS BETWEEN THE
DEEP TOW SURVEYS DEEMED IT NECESSA RY  TO MOVE THE MOUNDS SURVEY 150

FATHOMS TO THE NORTH.  SUSPECT TRANSPONDER SHIFTING IN THE RIDGE CREST
AREA DURING THE MONTH ON THE BOTTOM

                              TRUESDALE, CRANE
```

Telex sent from the Melville *to* Scripps *about the long-life transponders deployed at the Galápagos hot springs.*

as we slowly circled around it. By nightfall, our spotlights bounced off the transponder's reflector and the characteristic Scripps pink flag blowing in the breeze.

A cheer went up around the ship. This was my victory at sea. Dr. Spiess would be proud. Finally, the seventy-day expedition was over.

Russ and I flew from Panama City to Guatemala, where we boarded a World War II vintage plane that took us over the jungle to Tikal, in the middle of the country. There we cleansed ourselves of the sea, watched the flocks of parrots careen around the temples, and listened to the gurgling of the Montezuma's oropendola, a brilliant yellow-tailed, purple-bodied bird. We climbed the Temple to the Sun, hid from the tropical downpours, dodged the cascading water pouring over the Mayan steps, and lit candles inside our thatch hut. We shared stories with French and German trekkers, three Dutchmen, and some Americans. All the while, young Guatemalan children were taught English under the lantern lights. We saw spider monkeys, Central American deer, small warthogs,

Guatemalan wood hewers, fleet-footed vine creepers, and large pendulous bird nests swinging from huge trees. I treasured these moments, since I knew that when I returned to Scripps, I would face the dispiriting news that the discovery of underwater hot springs and the supporting data were no longer my province.

Indeed, Lonsdale and Weiss had drafted their article for publication in *Nature* while I was at sea. Many of my thesis data disappeared into that paper, and there was nothing I could do about the situation. There was not even a single line of credit acknowledging my work.

Shortly after Lonsdale and Weiss submitted their article, I decided that I would have a much greater chance for a future in oceanography if I worked more closely with Dick Von Herzen, my thesis co-advisor, at Woods Hole. I saw clearly that the salvation of my thesis did not lie at Scripps.

I left San Diego in the fall of 1976 and headed northeast across America into a world of bright fall color. Cindy Lee, one of my dearest friends, had graduated from Scripps and lived with her boyfriend on a landing craft in the middle of Great Harbor off the coast of Woods Hole. For one week I stayed with them, rowed ashore with them every morning, and rowed back to their houseboat every evening. By October 6, I had found a room in the house where Alice Cantelow lived. Alice had shared my cabin during the second leg of the *Pleiades* expedition. For a while, Woods Hole was a haven. I spent weeks peering at film, listening to the foghorns, and watching the hurricanes roll full force onto the Cape, with winds howling and trees bending. I assembled photo mosaics of the seafloor at the Galápagos Spreading Center from images gathered from the Deep-Tow and Woods Hole camera systems, which would be used to guide the upcoming Galápagos dives by the research submersible *Alvin* during the winter of 1977. During my last week in Woods Hole, I completed the geological map and geothermal guide to the Galápagos Spreading Center site, with all the hot springs and the long-life transponder positions marked, the faults mapped, and the volcanic structures and types keyed.

I also assembled a photo album that the *Alvin* divers would use to locate significant features on the ocean floor. In some photos, several white spheres appeared to float above the seafloor near the areas where we found temperature anomalies. Even though we could see their shadows on the basalt, biologists at Woods Hole insisted that they were nothing more than chemical mistakes caused by poor photo processing.

I knew that a man named Bob Ballard had inherited the surveying duties of the proposed 1977 *Alvin* expedition and that he would need a map to get to the site. Bob had never worked in the field of hydrothermal vents, but he had hundreds of hours of experience with the *Alvin,* mapping seamounts off the coast of Maine and at the Mid-Atlantic Ridge. The chief scientists of the Galápagos hot springs dives would be Jack Corliss from Oregon State University and Dick Von Herzen from Woods Hole.

Because I had a background that enabled me to help Ballard during the cruise, I went to visit him before I returned to San Diego. Bob was happy to meet me and especially interested in my map. "Maybe you would like to come along on this *Alvin* expedition?" he asked me. "Of course," I answered.

As it turned out, I was one of only two people from Scripps invited on the expedition. Ballard had his own reservations about Scripps, but I had the map to the hot springs and the keys to the Scripps transponders. It was my chance for a future, and without Bob Ballard's help, I am not sure that I would have had a career in oceanography at all.

I spent the Christmas of 1976 at my parents' home in Virginia. On Christmas Eve, my brother David, who had fallen sick in Colorado, was diagnosed with leukemia, and my family started its descent into a hellish nightmare. I didn't know what to do or how to help. I was ready to donate blood, bone marrow—anything—but there was little I could do. At my brother's insistence, I continued my thesis work, and I agreed to be the navigator for the return expedition to the Galápagos hot springs.

13

THE ALVIN
AND THE GALÁPAGOS

Paths curving, swirling back from the cliffs,
descending to the still abyss where cucumber masses
drift on warm thrusts . . . rising molten hot spurts
gentle pushes into time's eternity
—KATHLEEN CRANE, 86° WEST, GALÁPAGOS
SPREADING CENTER, FEBRUARY 1977

Outside, the red and orange winds off of Mexico were whipping the sea into a foaming frenzy. It was hot, really hot, and it was Christmas Day over the East Pacific Rise at 10° North in 1986. The French ship *Jean Charcot* was wending its way across the rise axis as it towed our instruments over the black smokers below. Commands echoed down from the bridge, a reminder of the celebrations soon to begin. Over on the starboard side, not even a mile away, a Soviet ship, the *Geolog Fersman* out of Leningrad, was mapping the same hot springs. Ten years earlier, we had discovered these springs at the Galápagos Spreading Center and, later, others at the East Pacific Rise. On the *Charcot*, I had no inkling that in another ten years, I would be sailing with those same Soviet scientists.

The French captain ordered that lights be strung up around the vessel. Champagne would be popped and fireworks would be shot

off. He wanted the Soviets to join in the Christmas festivities, and
so we radioed the *Fersman* and asked them in Russian to come and
join us. Surely, politics would fall to the wayside for holiday celebrations spent drifting more than 1,000 miles from shore.

The Soviets never responded; they continued to tow their instruments in search of black smokers on the seafloor. However, we
could see through our binoculars that many of the *Fersman's* scientists and crew were lining the railings watching our lights. I took a
moment to ponder how long it must have taken for the reports of
our 1976 discovery of hot springs on the seafloor to make it
through the Iron Curtain to the Soviet Union. Was it weeks? Years?
In the West, the first visit by humans to the Galápagos deepwater
vents in 1977 was reported like wildfire in the *New York Times* and
in the *San Francisco Chronicle*.

In the winter of 1977, still smarting from the turn of events following the Scripps hot spring discoveries, I prepared myself to participate on the first *Alvin* dives to the Galápagos Spreading Center.
Prior to departure, I spent a week with my brother David, who was
undergoing treatment at Georgetown University Hospital in Washington, D.C. After many long, deep talks with him, I wrapped up
my equipment and maps and headed to Washington National Airport, leaving my parents alone with their impending grief. I had a
job to do as the navigator for the first submersible expedition to the
Galápagos hot springs.

February 6, 1977
Washington National Airport.   I need to revamp my thoughts and
gear them toward the sea and the upcoming four weeks. I have
constant thoughts about David. During my stay in Washington, we
discussed such topics as the force of gravity. Are we as individuals
wrought together to maintain a form until the inertial acceleration

of a combination of stars alters us? What makes me think these things? It is David, dying.

5:45 P.M.   We are over Cuba now. I contacted Bob Ballard, Ken Green, Dick Von Herzen, Tom Crough, and the *National Geographic* photographers. They are engaged in the usual pre-shipboard talk: amounts of booze, will we get malaria, and do we have enough quinine? The *Lulu* and *Alvin* are late arriving into Panama due to engine problems in St. Croix. Still, we shall depart on schedule aboard the *Knorr,* drop transponders and perhaps test out the pressure cases before we return, pick up Jack Corliss, and make an equipment transfer. I'm to sleep on a cot somewhere in the ship because we have a real shortage of space. Perhaps I shall choose the "steel beach," if the weather holds and the stars are bright. I wish David good luck, while I am gone.

February 10

We are back in Panama after two days testing gear at sea. The Scripps transponder interrogator is now functioning. I hope the long-life transponders will, too, after all we went through last year. We have been using the navy multibeam maps, which are now declassified, to contour the seafloor at intervals of two fathoms. Bob and I are working out volcanic data solutions for the origin of the rift valley.

Back in Panama City, I went out to dinner at Casa del Mariscos with Emory Kristof, our photographer from *National Geographic* magazine, Bob Ballard, and the navy men. I was amused by their stories of underground dinners on the Azores (100 deep-fried canaries) as we ordered baby eels. The dinner bill totaled $91, and Emory took the tab. *National Geographic* pays for the bill only once, but they certainly make the "once" seem worthwhile.

The scientists are all geologists, most are married, and of course, most are men. I wonder again about marriage and family and the stability it must provide to my male colleagues. Many times I feel lonely or strange for not leaving anyone behind. Debbie, Terri, and I have been assigned to one cabin. We finally were given a place to stay below decks, which is amazing in its own way. I remember that up until recently women were never allowed below decks. I have an upper bunk again.

Today I continue work on the transponder positions and the Scripps interrogator. The mounds maps need some work. There is not much science to report. Dr. Von Herzen and Ken Green still haven't resolved the heat flow from last summer's *Pleiades* expedition, so I see the problems with navigation facing us again. But that is why I have been brought along: to find the sites. I have to get us back to the hot springs.

Emory and I keep ourselves amused and in shape by running around the fo'c'sle fifty times, about one mile. We reverse direction after the twenty-fifth lap to erase any ill effects induced by clockwise vorticity. It was on our fortieth lap today that this note, a folded piece of yellow lined paper, was delivered to me. It was from my friend, Scripps geophysicist Marcia McNutt. She must have been in Panama, too:

(Caustic) Kathy Crane
R/V *Knorr*
Dear Kathy,
This will have to be quick because we sail in 1/2 hour. Sorry we missed you! Have a good trip & stay out of trouble. If you see John Hadley, tell him I tried to look him up but 2 days in here wasn't long enough to track him down.
See you back at the mines.
Anchors aweigh!
Boom Boom McNutt

Marcia's note made me smile. She must be shipping out on the *Melville.* Having been well trained in explosives, she runs seismic refraction programs, tossing off charges wrapped in primacord from the fantail, while the next minute she'll return to the lab to polish her nails. I wish I had Marcia's self-assuredness to be a woman and to be utterly confident on the job. I feel that to fit into this submersible operation I have to become completely sexless, so that nobody will notice that I am any different from the others. In fact, I would like to be known as the best of the team, not the best woman on the team.

February 11

We're in Panama again, at the Rodman naval base, waiting for *Lulu.* I watched the birds with Jack Spiegelberg, a Scripps technician, while our pilot towed us into port. Bob Ballard and I walked to the post office and onto the U.S. Marine barracks past the heavenly flamboyant trees, the African tulip trees, and the monkey pod trees. He talked about his earlier work training horses, then porpoises and whales, free and easy in Hawaii. We spoke about the philosophy of oceanography and of those who make it their lives. We also talked about the stresses of going to sea, about how it takes about a month to prepare for a cruise and a month again to wind down from a cruise.

He and I dined with the *Knorr*'s captain and his wife, Von Herzen, and a Panamanian guest. We ordered enormous and wonderful seafood platters. It was the feast, perhaps, before the famine. We finished off the wine, played a wild game of Ping-Pong, and then gazed at the stars above and at the lurking fish and sharks underneath the lit pilings.

4:00 A.M.   Rendezvous with the *Lulu-Alvin*: Lights are glaring all over the ocean, piercing the darkness. This contraption is a rust bucket pontoon ship, holding up smiling weary faces. The transit through the Panama Canal locks was fascinating for the *Lulu* crew.

I long to be on the cruise with them. I hope the *Alvin* dive comes through for me.

February 12

11:30 P.M.   I just finished a large, two-fathom contour interval map of the spreading center. Tonight I showed the scientists the bottom photos from our Scripps *Pleiades II* 1976 expedition. The Woods Hole group is fascinated and enthusiastic. There is no love lost between the Woods Hole and Scripps scientists working on undersea hot springs, so it seems that my bad luck with scientists at Scripps has led to my good luck here.

February 13

Finally: I located the Scripps transponder Green III. After five hours of running around and trying different positions to ping on it, it awakened from its year-long sleep. This instrument, which I deployed into position with Bob Truesdale the previous year from the *Melville,* marked the hot spring we discovered, Clambake I. Don, the first mate, sensing the seriousness and the thrill of finding the transponder, yelled down from the bridge, "Fire Away Hot Lips!" I had no idea that this was the nickname that the *Knorr* crew had given me. I did not take offense. It gave everyone a big laugh.

February 14

I slept only one hour before it was time to cast the survey net. We homed in on the Scripps Green III transponder and set off from there. The new transponders will be named following Woods Hole tradition, after characters rather than colors. Ours will be called: Snow White, Sleepy, Dopey, Bashful, Grumpy, Doc, Sneezy, and Happy. In the spirit of naming characters, Bob, Emory, and I overheard the crew rename the three of us: The Good, The Bad, & The Ugly.

The Woods Hole ANGUS (Acoustically Navigated Geological
Underwater Survey) system

11:40 P.M.   We have put in another transponder. And the cooks
served ice cream *twice* in one day. I sit, listening in on the Woods
Hole transponders. After years of using the Scripps systems, with
their characteristic bubble-like sounds, I now hear belching and
gurgling sounds, like water drops rising from the depths. This is the
music of these new Woods Hole transponders, and they will be our
homing devices—the anchor pins for our upcoming dives. Bob
glanced by and palmed a candy bar into my hand while I was trying to fix in the position of Bashful. He knows how to train porpoises and people.

February 15

11:00 P.M.   We have just finished a 14-hour camera run with the
ANGUS (Acoustically Navigated Geological Underwater Survey)
photo system. Emory is dropping his bait trap and camera over

into the depression tonight, and Ray Davis will try for a rock core in the pahoehoe plains.

February 17

5:00 P.M. This has been a mentally exhausting day, correcting and recorrecting annotations of the first camera run. We placed three more transponders in the mounds area last night, and I got the Scripps Blue III transponder to answer beautifully, so now we can easily position our survey area off of the rise axis. After the survey, they lined up to within 100 meters of where we thought we put them. Good work Blue III—our 1976 Fourth of July baby.

Emory's "Easter Camera" was deployed. It is so named because it looks like a cross and is scheduled to rise on the third day. There was so much theatrical commotion over this ensemble that no one stopped to check the transponder attached to the camera rig. Bob emerged from the main lab with a glum look; the transponder was not working. After some time, the right jacks were connected, and we received the incoming signal. The *National Geographic* team tested the instrument all night letting the strobe flash while running film through the camera. Well, we have a first-of-its-kind film-processing van on board, allowing *National Geographic*'s Pete Petrone and Al Chandler to develop all the footage.

Yesterday, we had our first *Alvin-Lulu* rendezvous at sea, and the transfer resulted in three people going to the *Knorr* and three going to the *Lulu* (Jack Corliss, Jack Dymond, and John Edmond).

February 20

4:30 A.M. We have just finished a camera run and located one large temperature anomaly (0.1°C) with Ken's digital water temperature instrument. The anomaly was about 100 meters wide, and its position corresponds closely with Clambake II, located east

Clamshells photographed at the Galápagos hot springs, 1977. Woods
Hole Oceanographic Institution

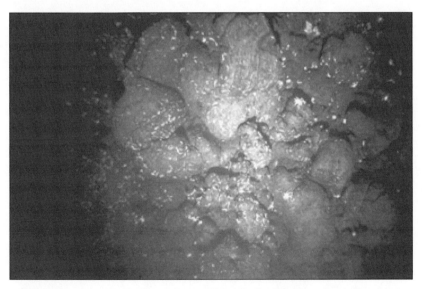

Serpulid worms and crabs at the Galápagos hot springs, 1977.
Woods Hole

of the underwater volcano, Mt. Swift. I also directed the *Alvin* to Clambake I and told the novice divers to be on the lookout for white clamshells, because they would lead them to the springs. It worked. The clams actually live in the warm water.

John Edmond just returned from his first dive. He was ebullient about the scene down below, with its shimmering water and huge white, dead clams (three species of pelecypods and shrimp and crabs).

The sea life was so luxuriant that the submersible's portholes were coated with all manner of creatures. The curious white spheres that we observed in last year's photos turned out to be animals, which we have named "Dandelions" (later identified as siphonophores, a relative of the jellyfish Portuguese Man-of-War). Other creatures have been found: beautiful tube worms (later identified as a new family of Vestimentifera), with scarlet and pink heads above pure white stalk-like tubes, and piles of "spaghetti-like" worms, which we as geologists called simply "Spaghetti" (later identified as enteropneusts, worm-like relatives of early systems of vertebrates). In all the enthusiasm, these first divers seemed to forget the scientists who paved the way, locating and mapping this exact area of the vast Pacific for their underwater excursion, including Von Herzen, Williams, Detrick, Klitgord, Lonsdale, Macdonald, Mudie, Weiss, and me.

February 20

3:30 P.M. *Alvin* found our Clambake II hot spring today, where there is a change in temperature of +5°C.

February 22

4:30 A.M. We had to abort camera run #5A due to pinger problems on the sled. Ken had reversed his instrument so that it would not fall out of his pressure case, yet we neglected to turn on its pinger. We had no bottom trace, and it would have been easy to crash the system into the seafloor. We brought it up, fixed it, and sent it down again by 1:00 A.M.

I am hoping that Bob secured a position for me on a dive the day after tomorrow. I'd like to see firsthand the shimmering water that we found last year. Emory and I viewed the divers' photos of the hot springs. The images showed that lovely clouded water in discrete forms, curling out like pudding over the rocks. Some areas were just gray and misty from the hot water.

Emory also loaded up his Nikon camera, gave me additional film, and sent me on a photo expedition around the ship—up to the bridge, to the captain's boat, everywhere. I need to learn something about this camera if I am going to document an upcoming *Alvin* dive.

February 22

8:00 A.M. Finished camera run #5. We had a fun-filled night with popcorn, music, and a convivial mood

The author, with Bob Ballard, performing in the Equator-crossing ceremony to celebrate the first Alvin dives on the Galápagos vents, 1977
PHOTO: *Emory Kristof*

shared by all. Emory is in charge of the traditional Pollywog skit, which will be performed when we cross the Equator. The Equator crossing party will include an *Alvin-Lulu* parody with Corliss and Ballard as the *Alvin*'s chief scientists. The production will be similar to the movie *The Neptune Factor*, featuring hydrothermal vents, giant creatures, and mad scientists. I am to be the queen of the court. The crew will enter two characters to the Equator Crossing Royal Court.

I am scheduled to dive tomorrow from the center of the rift axis to the northern inner wall.

February 23

Dive 719.   The sole objective of *Lulu*'s crew is to get the *Alvin* into and out of the water safely. After 9:30 P.M., the submersible shop closes down until 6:00 A.M., when most of the crew rises to set up the navigation net and to review the checklist for the next dive. The names of the crew are really funny when taken together: Brody, Foo, Panama, Porty, Jack the Whaler, Captain Pflegenheimer, Cook, Dudley the pilot, Jack Donnelley the pilot, and John Jain.

Last night, I took a Boston whaler from the *Knorr* over to the *Lulu* on glassy seas. We use the whaler to transport people and samples from the sub dives back to the main labs on the *Knorr*. Once aboard the *Lulu*, I was given a bunk in the navigation van because women are not allowed into the tubes, where the crew members live. Van Andel gave us a briefing about the previous work, and Jack Donnelley gave a briefing in the submarine to make sure that we would not suffer from claustrophobia. Evening on the deck (there is no where else to go on the giant catamaran) was spent looking up at the stars and into the ocean. Women were allowed only minutes in the shower since both the bathroom and the galley open out onto the pontoons where

Alvin *prepares to dive in the Galápagos Spreading Center, 1977.*

there is a free and steady flow of people. Emory calls this ship the biggest flush toilet in the world.

More so than others, this crew does not like women on board their ship, and it makes me more uncomfortable than usual. Breakfast was nothing but a greasy omelet, and I had to cease all fluids because I did not want to use the HERE (Human Range Extender) bottles in the sub. I heard that some of these guys drink so much coffee before a dive that they completely fill up all the bottles within one hour of submergence.

I put all my gear in my dive box. My diving vest was filled with pens, papers, camera lenses, navigating tools small enough to use in this very tiny space, my watch cap for those freezing hours on the bottom, and warm clothes. The captain gave the command: "Prepare to launch the whaler."

Crew members grabbed the ropes, Ralph manned the invisible capstan, we climbed in, and the hatch was shut. I crouched in my corner on the port side—Bob advised me that this would be the best view. John Edmond, with his very long legs, was on the star-board. First, the *Alvin* was extracted from the innards of the *Lulu,* and we were pulled to sea like a bobbing cork. All I could see was a dimming blue light, bubbles rising upward, and a sudden stilling of the waters all around us. We were sinking. Down we went, with lights illuminating the deep-scattering layer. I could never have dreamt of the beautiful fish or long train-like gelatinous substances that glowed off and on, of the darting arrows or the crab-like tiny beasts. It was a light show of the most startling and gentle colors. Then we slowed, and our pilot Donnelley peered out his forward port. "Bottom in sight," he said.

The view was eerie, a darkness looming from what seemed to be the darkest place on Earth. This alien blackness was deepened by the fearful emotions generated by looking into a forbidden world. We crawled along, spotting empty lava tubes and giant craters and lava drain back holes. These appeared to be caverns into the very heart of the volcano.

Here and there, however, increasing with every foot we moved, the black seafloor started to glitter with a thousand splinters of black, sharp light. We were entering the fields of volcanic glass, tor-tured and crystalline, black mirrors reflecting the first glimpses of light ever to reach so far down. Volcanic glass meant that the erup-tion had been recent; otherwise the chemicals in the sea would have weathered all the shiny glass facets and turrets into forms of iron oxide, leaving nothing but dull lumps of seafloor pillow basalt.

Suddenly, we saw an even more mesmerizing display. We entered warm, azure-colored ponds of life isolated in the hostile, extreme darkness. And there were the clamshells—huge, white, and wedged in the cracks between the boulders of pillow basalt. They basked in the warm spring water that oozed upward into the ocean above.

Emory Kristof, Dudley Foster, and Kathy Crane on the Galápagos Spreading Center, 1979. Woods Hole

The temperature here was 14°C, almost balmy compared to the seafloor temperatures of around 2°C.

It amazed me that this was the place where I had launched the Scripps transponder a year earlier. Our perceptions change when we see for ourselves the remotest parts of our world, and we must change our fundamental notions of the creation of the seafloor, of the oceans, of life. Amidst this alien landscape, we took samples. Although we are geologists, I am culturing bacteria for Victor Vidal, a graduate student at Scripps, who believes that life thrives on the deep-sea springs. He has instructed me how to sample bacterial slime and how to put it in seawater with 2% gluteraldehyde under sterile conditions. All the samples need to be incubated at 60°C, and I have been trying to convince the divers to gather more samples.

The discovery that most of the seafloor is covered by sheet-like flows of lava rather than pillow basalts (which have been photographed in the Atlantic) is tremendous. Bob told me that only the previous month, some very eminent geologists submitted a scientific manuscript for publication explaining in great detail why sheet-like lava flows could never erupt onto the seafloor. I think our findings will cause them to withdraw their submission.

In the case of the Atlantic, the seafloor grows very slowly, so sediment trickling down through the water column can accumulate on the ocean floor. Sheet-like lava flows lie more horizontally and thus are covered up more readily than the flows that erupt into pillows and form the central high volcanic centers in the Mid-Atlantic rift valley. In contrast, the Galápagos Spreading Center spreads more rapidly along the Pacific Ocean floor, allowing the sheet flows to be free from the accumulation of sediment layers.

And at last, we have visual proof for the hot springs. They are real and not some figment of imagination. How will this affect the long-standing oceanographic models of heat flow and ocean currents? At Scripps, Joe Reid made a deal with me. "Okay, Kathy," he told me, "you can disturb the bottom of the ocean but leave the top to us."

In fact, the models for ocean circulation are still based on a horizontally stratified sea, with the densest, coldest water on the bottom. In these models, the heat that drives the system comes only from the sun. Instead, we have found massive quantities of heat rising from the seafloor below, from hot springs, just like the one we were sitting beside in the submarine.

Our goal on this dive is to make a beeline north to the marginal high of the rift valley. This is challenging because navigation is problematic in the sub. I knew how hard it was to reconstruct the navigation for other people's dives when they had not recorded accurate notes about their positions. Many of the places they discovered remain in their tortuous submarine pathways. If we want to reconstruct where we go, then all the divers need to document everything, especially heading and position coordinates.

We left the central high and headed north. The alien darkness returned, with its oppressive weight. We reached the north wall, which we were going to transect in order to map the percentage of pillow flows compared to sheet flows. Escarpments of faults are slices through geological time, and one could map the episodes of volcanism by looking at the sequences of these flows. How often did eruptions take place? What kind of eruptions were they?

It was late by then, lunchtime, and we pulled out our food box, a giant binocular case filled with roast beef and ham and cheese sandwiches, and tarts for dessert, with coffee and juice. By this time, my knees, which were crumpled up beneath me in a fetal position, screamed in pain. John Edmond's problems were worse with his long legs—he just simply stuck his feet in my porthole. Now I had even less space.

We sampled rocks again, storing them in their proper containers, took pictures, video tapes, and notes, and recorded fault orientation by looking at the CTFM (Continuous Transmission Frequency Modulated) sonar. The CTFM sonar creates 360° images, and in the submersible it is used for obstacle avoidance. We also reestablished our communication with the *Lulu* navigation control, somewhere thousands of feet above. The radio echoed over and over, burbling and gurgling up through the water column. It was time to surface. We put on a tape of music and rose, over several hours, to the top of the sea. We could feel the surface before we could see it. The sub took a surge downward and then another

Alvin *returns home to the Lulu, 1977.*

sickening surge upward. We were flotsam caught in the power of
the waves.

The *Alvin* pilot assumed another set of communications with
the rescue boat and its divers, and a collar was attached to the sub.
We were towed into the *Lulu,* and it became very quiet in the sub.
Finally, the hatch opened, and we could see the blue tropical sky
above—a beautiful, welcome sight. I was the first one out, and I
could barely straighten my legs. Everyone cheered and cleared a
pathway to let me run straight away to the bathroom. Then came
a huge dinner for the divers—eggs, fried rice, steak—and then
blessed sleep at 10:00 P.M. in the navigation van. The *Knorr*'s radio
communications kept waking me until I located the volume control for channel 6 and turned it down as far as it would go in order
to get much needed rest. The next morning, I ate breakfast,

worked through the navigation, sat in the sun, read transcripts of the dive logs, watched another launch of the sub, and took a swim in the cradle of the *Lulu* during the *Alvin's* absence. A sea turtle sailed by underneath our feet. The thought that there were 2,500 meters of water beneath us was awesome. By evening, it was time to rescue the *Alvin,* and it was my job to ride in the Boston whaler with the divers. My position was "bow hook." We saw the red sail of the *Alvin* and raced out to meet it. The divers jumped into the sea, drained the sail, and then we attached the sub to the whaler and towed it back to the *Lulu.* After recovery, all the samples— sediments, rocks, water, and biota—were collected and put into freezer boxes. The coolers and I were lowered into the whaler readied for transfer back to the *Knorr,* where I was greeted by Morse code in lights and a huge sign painted on several bed sheets spread out over the bridge: "Welcome back Kathy, who did it deepest."

February 27

I have just heard that we'll put into three different islands in the Galápagos. Each night we will go into a different bay. I so look forward to just tramping about on land. I have had such a yearning to see the Galápagos that I nearly forgot to mention that another Clambake was found today. It is located along the western central high scarp, where the fissures cut through the terrain.

February 28

Today we cross the Equator.

I awoke at 2:15 A.M. and took over the 2:30–7:00 A.M. shift. What pleasure it is to manipulate the camera over the choicest spots of the ridge crest, maneuvering like a slalom skier over the terrain. Earl Grant (Earl the Pearl) and I worked together, and I took extra caution to maneuver the camera sled over Clambake III. Bob showed me the processed color photos we took when we deployed the ANGUS to let me know that I had successfully flown the sled over the spring.

Bob Ballard and Emory Kristof have allowed me so much responsibility and have given me so many words of support.

Now I have a decision to make. Should I stay here on board for the next leg? It is exciting. We are making history and publicly this time. Yet every day, every hour, some part of my brain is thinking of my brother David. I must go back to him.

Tomorrow we arrive at the Galápagos.

March 2

Yesterday, we reached San Cristóbal and Wreck Bay early in the morning, but we were cleared for shore hours later, at 11:30 A.M. We had to anchor out in the harbor next to *Lulu* and a confiscated Japanese fishing boat, *Niku Maru #21*. Three armed Ecuadorian guards paced on the docks. Crew members fished over the stern. We packed our masks, snorkels, and long pants, intent upon making it to the pier as one unit. Bob, Emory, and I landed together and hiked up into the scrub of the island to a road about five hot, dusty miles away. Along the way, we encountered yellow finches, black finches, a few lizards—some with red bellies—a dead owl, a stand of lime trees, and one strange legume tree with clusters of red fruit. We made it up to the road to the red cliff and turned back toward the town to get *cervesas* (beers) in the local tavern where some of the *Knorr's* crew members were busy getting drunk. I chose to walk to the restaurant across the street where Emory purchased the hulking brown bottles of frothing warm Ecuadorian beer. Outside on the southern peninsula, marine iguanas, large slate-gray beggars, nosed for scraps of food.

Lizard trails intertwined endlessly over what little sand dusted the volcanic rock.

Since there was no beach, we clambered over what seemed interminable miles of lava boulders. Each step grew harder and harder, and I had that strange feeling that I would collapse from sunstroke. Fortunately, the torture ended at exactly the moment when we charged into the ocean. The swim saved us. The feeling was luxuriant as we plunged into the cool, massaging water. Butterfly fish, a giant parrot fish, spiny urchins, a school of some silver fish—like herring—played among the boulders beneath us.

Later that evening, back on the ship, I had a long talk with Emory about my future and about my brother. These two issues vied for space in my mind, and I found no solace. Later still, Bob and I went up to the bow to stick our head out of the bow ports, whistling and watching porpoises as they dodged in and out, arching away, breaching, and whistling back at us. Bob knew all their codes. I am sure that they were whistling to him.

March 3

Today we stopped at Floreana Island to visit the post office and to get our cards and letters stamped. As we did the previous day, we took advantage of being on dry land.

From Floreana, Bob and I backpacked to a more remote location on the island where the pahoehoe lava flows were coated with a gentle dusting of sand, and cactus trees were in bloom. Pacific waves crashed and foamed against the lava pilings. While sea lions frolicked in the waves, we spied two seal pups basking in the sun. There we took pictures of the creatures: penguins, small and black, and pelicans, darting around their nests on top of thick bushes, poking their heads above the masses of leaves to observe strangers. To dive in the water, to feel the cool glide of liquid over my limbs, and to swim alongside the seals in their underwater world was a delight. The island across from us looked like one large pahoehoe blister, nearly identical to the kind that formed the undersea caverns we just explored on the *Alvin* dives. The island is covered by the oddest assortment of flora and fauna: mangroves and cactus, a sea lion rookery, bodies lounging everywhere. Pelicans and brightly colored seashells collected across the pink-white sand, and all the while, the susurrus of the light blue water, up and back along the scimitar-shaped beach.

March 4

I am over the Pacific Ocean between the Galápagos Islands and Ecuador, where I shall make the arduous journey by bus to the Quito airport.

I am thinking of these last days after the *Lulu-Alvin* expedition. After Floreana, we had a spectacular evening. Back on the *Knorr*, we gazed at the stars and the nearly full moon while we warmed ourselves next to the smokestacks on the bridge deck. Earlier, Emory commandeered the last eight lobsters on Santa Cruz island, home of the Darwin Research Center, and Bob and I roused ourselves to head over for dinner.

On Santa Cruz, we sauntered around the open-air houses, past the imported palms and hibiscus, where we came across the small hotel overlooking the lava boulder–bordered waterfront. There we found MIT geophysicist Tanya Atwater, writing and revising a National Science Foundation proposal. She had come to join the second leg of our expedition, to prepare her for future dives in the *Alvin* to the Mid-Atlantic Ridge. Tanya, tired of proposals, joined Bob and me, and we paraded eastward toward the Darwin Station to view the tortoises.

Dinner was at 8:00, with lobsters and Chilean wine, alongside the moonlit water. The shimmering ripples lapped gently over the weathered black shiny pebbles that washed up between the outward jutting basalt. Dinghies pulled lightly at their lines, rotating slightly this way, then that.

It is strange how circumstances can bring people together. I wondered, in contrast to the *Pleiades* expedition, why should *this* group have gotten along so well together? We boarded the launch at the pier, singing "We All Live in a Yellow Submarine," as sailboats glided by. We pulled up to the *Beagle,* a sailboat owned by the Darwin Station where one of our crew dismounted and joined the rollicking crew on board. From boat to boat, we put-putted out to the *Knorr.* Once we were alongside, we climbed onto the ship. Later, outside of Bob's cabin, we decided to make good on our nickname, and we played the soundtrack to *The Good, the Bad, and the Ugly* and watched the languid moon over the sea. I shall miss those nights.

When we are at sea, we are away from our pasts and away from our futures, away from our land lives and loves. When we go home, we inevitably fall back into our previous roles. My expedition friends will return to their wives and families. I will rush to my brother's bedside.

For a time, when we are sailing together, we are each other's family, as real as any on this earth.

March 5
Quito Airport.   It is over. I am going home.

The events of the Galápagos hot spring dives represented some of the best moments of my life—and some of the worst. We had made fantastic underwater discoveries, but soon I would face the death of my brother. All the joy and chaos on board the *Alvin* and *Lulu* paled somehow when I thought of David, hanging onto the last threads of his life.

I returned to Washington and from there to California on March 10, in time for my sister's birthday. It seemed that everyone was excited about our hot spring discoveries. Walter Sullivan wrote a *New York Times* article featuring the vast implications of this find, and this single article had more power to move the scientific community than any research scientist–authored journal manuscript written in the conventional mode:

Hot Springs on Ocean Floor Found Teeming with Life
by Walter Sullivan (excerpted)

The first manned exploration of active hot springs on the deep ocean floor has found that some of them are focal points of teeming life in an otherwise largely sterile environment.

An account of the 25 dives, conducted in as many days, was presented Wednesday evening, by Dr. Robert D. Ballard of the Woods Hole Oceanographic Institution at a meeting of the Woods Hole Associates at the New York Yacht Club.

The dives were conducted to depths of about 9,000 feet in the Galápagos Rift Zone by the Woods Hole submersible *Alvin* supported by two other of the institution's vessels: the submarine's mother ship *Lulu* and the research vessel *Knorr*.

The submarine carries a pilot and two scientists. Among the 13 scientists who made dives in the expedition, which ended March 24, were three women. Two were graduate students, Kathleen Crane of the Scripps Institution of Oceanography at La Jolla, Calif., and Deborah Stakes of Oregon State University in Corvallis. The third was Dr. Tanya Atwater of the Massachusetts Institute of Technology.

National Geographic and the *New York Times* were our marketers, our judge and jury, and that combined clout was astonishing and effective. Every high-powered biologist wanted the seafloor samples we collected. I was the only one on board the *Knorr* who tried to culture thermophyllic bacteria as requested by fellow graduate students at Scripps. When radio reports relayed the biological find to Woods Hole, microbiologists there ordered us to stop everything and concentrate on sampling more biota. Suddenly the biological culturing I conducted seemed very important. Since I was heading to the United States to see my brother after the first leg of the expedition, I left the cultures in the hands of other scientists on board the *Knorr*. After the second leg, the samples never made it back to San Diego; they were "lost in transit" in Miami.

My heart went out to Victor Vidal at Scripps. I understood his profound disappointment, and I felt as though I had let him down. At Scripps and Woods Hole the smell of competition and greed was rife among the research scientists; all those chemists, biologists, and geologists were now converted believers in deep-ocean hot springs. While it was patently clear that we had discovered new life-forms, they all wanted to claim a piece of the work and call it their own.

Many other articles were written about the discoveries and about the three women who had descended to the hot springs on the seafloor. Debbie Stakes from Oregon State University and Tanya Atwater from MIT had descended in the *Alvin* in the second leg of the expedition. The *Washington Post,* the *Los Angeles Times,* the *San Diego Union,* many foreign journals, and even *Fish and Wildlife* called to interview me, but I stalled many of them. My thoughts and emotions were racing in too many directions at once. The strongest feelings were those concerning David—I desperately wanted to be by his side. I talked with Dr. Spiess about leaving Scripps so that I could spend time with my brother. I remember crying, feeling so embarrassed, but feeling utterly broken, too. In his quiet way, Dr. Spiess told me that, yes, he understood. After all, he was the father of five children.

However, I did not leave Scripps; David had asked me to stay.

My concentration diminished. In my despair, I threw myself into dangerous adventures to reaffirm my own sense of life. I camped in Joshua Tree National Park and nearly froze just to look at the stars, just to examine the exquisite wildflowers under a crust of snow. I went to Death Valley and walked tens of miles, in and out of the heat, through the hills and through the hell of my mind.

One day I walked to Scripps along the coastline, completely neglecting my training as both a water safety instructor and an oceanographer. I got caught in a high tide. All the coastal pools were submerged by the tide, and water roiled in and out. I was so determined to make it to work that I shinned up onto a narrow surfer's trail along the cliffs, with my draft thesis in my backpack. Masses of waves crashed against the cliff, except for one ledge covered with slimy mud about eight inches wide. To get to the ledge I had to build a three-foot-high hill. It took me half an hour to find enough cobbles, and after three tries, I succeeded finally in hoisting myself up onto it. Below me were treacherous sea caves, and there was no way back. I moved around the corner, and the entire ledge dropped away into the ocean. Only the bare cliffs remained. I took off my pants and shoes and put them in my backpack with my thesis and climbed down into the water when it ebbed. That sea was a powerful and monstrous beast. I made it to the first headland when the waves surged over me. "Don't let go! Don't let go," I kept thinking. "Don't let go! Don't let go." I chanted this over and over until I made it to shallow water, and then I rushed out of the ocean up to Scripps, panting, gasping, exhausted. I was soaked to the bone, so I stretched all of my clothes on the grass to dry, while I dressed in the spare set I kept at my office.

Larry Mayer was walking by the pier when he saw me dash bedraggled from the sea with my backpack held over my head. He shouted, "Kathy, why did you come to work this way! Are you crazy or something?"

I certainly must have appeared crazy. I would write in the morning and then take walks for miles and miles, trying to escape from the turmoil in my mind and in my heart. On April 29, I telephoned my family in Virginia. It was David's birthday. He couldn't even crawl out of bed. On May 3, 1977, David died. He was twenty-seven years old.

> When the Dark comes rising,
> Six shall turn it back,
> Three from the circle,
> Three from the track,
> Wood, Bronze, Iron;
> Water, Fire, Stone;
> Five will return,
> And one go alone.
> —S. COOPER

My siblings had been six and now we were five. We buried some of my brother's ashes in a Virginia churchyard. Afterward, my parents and I also took him back to his Colorado cabin and threw his ashes into the wind. David was free.

Shortly thereafter, I returned to Scripps to continue work on my thesis. I felt emotionally shattered and, perhaps impulsively, I did two things. I sent an application to NASA to enter the competition to become a space shuttle astronaut, and I signed up for the next ocean expedition, scheduled to leave Panama on May 30.

I can hardly remember where we went on that expedition or why. We were working in the Pacific Ocean somewhere, and it felt like nowhere. Instead of tropical flowers, I saw only shantytowns. Instead of vanilla, I smelled urine and festering garbage. Instead of laughter, I heard only children whimpering, their mothers crying. I remember full moons off of port bows, swelling waves, tropical rains, passing by the Panamanian prison island, Coiba, with its steep cliffs to the south and dark, dome-shaped hills, surrounded by crashing waves. When I managed to sleep, my dreams were always of death.

I tried not to sleep and to simply survive. Two passages by Pearl Buck described this voyage of mine, a journey through darkness that led to a glimmer of hope, to a new vision of life and how I should live it:

> To the sea I go with love and terror,
> for actually I am afraid of water,
> and I'm never deceived by calm under sunshine
> or even under the moon.
> The madness is there, hidden in the depths of
>   unknown caverns. And yet I go back to the
> sea again and again.

I believe in family, ancestors and all. The individual is a lonely creature otherwise in this changing world.

14

SPACE CADETS

Nothing of him that doth fade,
But doth suffer a sea change
Into something rich and strange.
Sea Nymphs hourly ring his knell:
Hark! now I hear them—
Ding-dong, bell

—SHAKESPEARE, THE TEMPEST

After my brother died, I lost interest in a future of pure research. However, I somehow went through the motions of being a scientist: I coauthored papers about our hydrothermal discoveries, and I knew that I would defend my Ph.D. thesis in September. This document presented the bulk of my Galápagos Spreading Center work in addition to the early results from the East Pacific Rise and the Reykjanes Ridge.

After I finished writing my dissertation in August, I went to an international conference in Durham, England, to present some of our hydrothermal results. On the first day of the conference, my friends and I rented a rowboat to see Durham by water. A boat filled with Soviet scientists, who were attending the same conference, floated next to us. I was eager to meet my counterparts from behind the Iron Curtain, and we exchanged resounding "hello's." Soon we were racing to the next bridge. I don't know who won, but the spirited race was peppered with Russian and American laughter. At the finish, we rowed close to each other and shook hands. I introduced myself to Elena Lubimova, a heat-flow expert from the Shirshov Institute of Oceanology in Moscow. This marked my first meeting with a Russian scientist. Elena and I had read each other's research papers, so we were very excited to meet. The other scientists in her boat had thought that K. Crane was an old man and were surprised to learn that a twenty-five-year-old woman was the author of those papers.

After the conference, I went to Ireland to be alone and to hike. From there, I returned to California, where I put the final touches on my dissertation, *Hydrothermal Activity and Near Axis Structure at Mid-Ocean Spreading Centers.* I defended it on September 17. The conference room at the Institute of Geophysics and Planetary Physics was packed. Many of my former roommates sat on the floor. My Scripps colleagues lined the hallways to listen to the account of the discoveries made at the Galápagos Spreading Center. My committee sat in the front: Dr. Fred Spiess, Dr. Richard Von Herzen, Dr. Jerry Winterer, Professor Joe Reid, and Dr. Robert Luginanni. At the end of my hour-long presentation, the audience was invited to leave the room so that the committee could begin its questions. Truthfully, I don't remember what I was asked, but I must have responded well enough, because every committee member shook my hand and presented me with the precious "signed signature page," the document that would confirm my doctorate degree. All in all, I was awed and humbled by the huge crowd that attended my thesis defense. Afterward, even my parents showed up for the celebration, which was a rollicking party at Dr. Spiess's wonderful house. We danced, drank margaritas and champagne, and ate vast quantities of Mexican food. The next day, I began my postdoctoral work at Scripps, joined a row-ing team in San Diego, and began enjoying all the beauties of the La Jolla environment.

In October, my telephone began to ring and kept on ringing. Larry Mayer took the call, finally. I can still picture him, rapturous, bubbling with joy and awe: "Guess who called today, guess! They want us, Kathy, *us,* in Houston!"

NASA informed Larry and me that we were finalists in the new space shuttle astronauts program, which now included women. We were instructed to go to the Johnson Space Flight Center near Houston for one week of interviews and tests. Soon the word spread through San Diego, our institute, our families, to the other labs, and to the ships at sea. We were among the 200 finalists chosen from 8,000 applicants, along with Brian Shoemaker from the U.S. Navy, Joan Fitzpatrick from the Colorado School of Mines, Kathy Sullivan from the Canadian Bedford Institute of Oceanography, and Sally Ride from Stanford University. We had had some inkling of our good standing because one month earlier, FBI agents had been swarming around Scripps asking questions about our morality, ethics, and patriotism. My housemate Bob Kleinberg, a physicist at UCSD, had had fun explaining my relationship with five male housemates to the FBI: I ate the dinners they cooked. After the list of finalists was announced, local West Coast news reporters descended upon us.

When we arrived at NASA in Houston, we found sealed envelopes on the dressers in our hotel rooms. The envelopes contained instructions for the following day: Meet with the NASA astronauts and await further instructions. The letters of introduction advised us, in understated terms, that the job "involved considerable travel, overtime, and was not entirely risk free."

NASA had arranged to conduct the interviews in small groups of twenty, and when our group assembled the next day at the space flight center, each of us was assigned a blue bag that we had to carry with us everywhere. These would act as repositories into which we would collect our bodily fluids for one week, to be analyzed later by NASA physicians. They were not to leave our sides. We shuffled from one lecture and test to another, carrying our ubiquitous blue bags, as NASA employees chuckled at the parade.

During one test, NASA examiners put each of us into a tiny dark sphere, alone, where they monitored our reactions. The spheres were supposed to simulate survival capsules for the space shuttle. If an accident occurred, the astronauts would escape from one shuttle to another, inside these bouncing balls. One member of our group, another woman, fell asleep, claiming that every mother could use such an isolation chamber.

We also participated in the simulator rides. Commander John Young, who would become the first pilot of the shuttle, seated Larry and me in the seats behind him as he started the simulator's computer sequences. Larry and I watched in fascination. The computer directed the simulator to cause a malfunction in orbit, and during the simulated re-entry to Earth's atmosphere, we began to spin out of control. Commander Young kept trying to pull us out of a nosedive into Earth, but as the computerized California desert grew closer and closer, we were spinning faster and faster. In the flash of a second, it was all over. There were beads of sweat on his brow as an "X" appeared on our computer screen. Commander Young turned around to Larry and me and said, "We're all dead. Damn, I still can't fly this thing!"

In another meeting, we were called one by one to meet with NASA astronauts. They were seated at tables arranged in a C-shaped pattern, while the interviewee sat in the middle, with his or her back to many of the astronauts. We assumed that this was some kind of psychological test because it was uncomfortable trying to answer questions asked by people we couldn't see. I was surprised that the line of questioning encompassed issues of national defense, such as, "How would you feel about launching a Department of Defense spy satellite?"

I found out later that NASA and the military were working together to plan shuttle missions, since the federal government lacked sufficient funds to finance separate space shuttle programs for both NASA and the Department of Defense. One of my San Diego housemates, Richard Harms, an astrophysicist, explained that this kind of question was technically illegal, because NASA was interviewing *civilian* astronauts. It was obvious that NASA needed to cull any of us who were considered "anti-military." The Cold War just would not leave us alone.

The final test required each of us to run on a treadmill until we exceeded our aerobic capacity and went anaerobic. Few women had ever been tested. I remember that electrodes were pasted all over me, and a mask covered my face. I had to run for what seemed like forever while scientists in white lab coats—just like in the movies—scurried around, monitoring my carbon dioxide levels. Larry and I had a plane to catch, and he kept peeking through the door mouthing the words, "Go anaerobic, Kathy!"

When it finally happened, I heard the scientist scream, "She went anaerobic!" I felt like I was drowning in my fluids as I hopped off the treadmill, ripped off my robe, peeled off all the electrodes I could find, jumped into my clothes, grabbed my bags, and dashed out of the room. On the flight back to San Diego, I found two electrodes still stuck to my body. In a funny coincidence, Jack Donovan, the spelunker from the Deep-Tow Group, passed by us in the aisle. He asked us what we had been doing together in Houston, and we told him that we were astronaut finalists. He just stared at us and laughed. "I don't believe this! Larry and Kathy, astronauts? What has the United States come to?"

A letter to the editor that ran in the *Los Angeles Times* drew the interest of my housemate Bob Kleinberg. Written by Sarah Tamor of Santa Monica, it was titled, "Woman Astronaut." Ms. Tamor wrote:

When I picked up *The Times* (Aug. 30) and saw the front-page picture of Dr. Margaret Seddon testing to become one of the first women astronauts, I was delighted.

I could not have been more disgusted by the reporting that followed. The first sentence landed with a thud: "Romance in Space?" How infuriating!

This is 1977. It's about time that women's activities in the world be viewed in the same asexual light we take for granted in discussing men's activities.

Men journey into space and we talk about progress, technology, new worlds. Women journey into space and we talk about their "shiny blonde hair, shapely legs, T-shirts, short shorts, and blue eye shadow."

Bob had ticked off each of the items mentioned in the last sentence, except the last one, in a copy of the article he gave to me. He had added, "Memo—get Kathy some blue eye shadow."

Larry and I went home to San Diego, leaving Brian, Joan, and the sixteen others in our group to their own destinies. On our return, we were immersed in an atmosphere of indecision and waiting, distracted by memories of spheres, simulators, sound chambers, and the thrill of possibility.

Meanwhile, a new fear nagged at me, in addition to my recurring bugaboo: Could I really fit in with NASA's expectations? Surely I was made for this job, and I would do it well. In fact, I would thrive on it. But what about the other half of my life, if indeed it really comprised half? I now wondered about the part of me that I had suppressed for so many years, the part that wanted a home, loved poetry and music, and yearned for a quiet life of rural solitude, removed from the analytical, impersonal demands of modern science.

Nonetheless, I accepted in the meantime a postdoctoral position at Woods Hole to work with Bob Ballard, and I ran away to the East Coast in December 1977. The world turned cold and dark, and black and white, in that small Massachusetts ocean town. During the dark December, I kept wondering, "Why don't they call? When will NASA decide?" I felt as though NASA was keeping me hanging by a thread.

I remember the day—it was a Wednesday—when the United Press International (UPI) news agency phoned me at my Woods Hole office: "Dr. Crane, you are listed as one of the forty remaining candidates. Your chances have been upped to 50 percent. How do you feel, Dr. Crane, to be one of forty finalists out of 8,000 initial applicants?" I couldn't sleep that night. During the day, I heard people mutter, "Looks like she made it, by God. Looks like she made it."

I went away the following weekend because I feared the next telephone call. With so much expectation, I feared rejection. When I returned to my office, I discovered that NASA had been trying to reach me, and my panic increased. Instead of returning the call, I turned on the radio and listened to the news broadcast; it reported that six women astronauts had been selected for the space shuttle program. The next phone call was from the local Cape Cod television news station. "Dr. Crane, Channel 5 News *still* wants to talk with you." I knew then that I was not one of the six women chosen to become astronauts.

Shortly thereafter, I received a Telex. NASA had rejected me for medical reasons, and to this day, I do not know what the medical problems were that kept me from going into space. All I could think of was that I had failed. The phones stopped ringing.

My friends stood by me, however. Emory Kristof wrote to me that now NASA would be deprived of the ultimate free-fall ice cream–eating experiment and, recalling the *Alvin* expedition, added that Neptune's Court would be honored to retain me as its queen. Larry cheered me up with the report that nobody in our group of twenty had been selected. Fellow finalist Brian Shoemaker called to tell me about his rejection. He told me that he had assumed that he had an inside track, since his admiral had assured him that NASA would select him, and that in preparation, the navy had removed his name from the list of active military personnel.

I was very disappointed, and in response, I focused all my energy on my work at Woods Hole. I needed to eliminate emotional distractions, and so I chose to work on routine tasks. Together with Bob Ballard, I dove to the Mid-Atlantic Ridge and revisited the Galápagos vents. To reignite enthusiasm in science, I dashed off on a trip to Africa to map the pattern of hot spring and volcanic activity in the great East African Rift, considered to be a visible template of the more evolved volcanic spreading centers, which lay hidden under the deep ocean.

For two straight years, I had kept my distance from those scientists who were battling over the hydrothermal vent discoveries we had made in the Pacific in 1976 and 1977. It turned out that Africa was not distant enough, because some of those same people turned up there to do research. Even Harmon Craig and his entourage from Scripps converged in Kenya, and we ended up having to work side by side again. My field associate, Suzanne O'Connell, knew how much I needed to stay away from controversy, and she staunchly defended me during an entire month in the African dust while we were on assignment for the Research Council of the National Geographic Society.

The anguish of my brother's death would not leave me—not in the hot bush of Kenya, following the Masai with their scarlet robes billowing under the searing sun, nor under the ocean, sealed in a submarine, searching for signs of volcanic activity. I found no peace. I found no comfort. I felt distanced from humanity, and so I disappeared inside of myself—and my career nearly disappeared, too. Were it not for a few valiant souls who pressed me to stay in science, I might have vanished altogether. Despite the excruciating reality of the world, I learned to believe in the romance of scientific pursuit and discovery.

15

WOODS HOLE

Woods Hole: So much has been written about the beauty and magic of this small, world-renowned oceanographic center. After Scripps and San Diego, I felt rather alone, and I became aware of how conservative the people were. When I took the Woods Hole postdoctoral position, I was very young, just twenty-six, and I was very conscious of my gender as part of the staff of mostly older men. I encountered some very strange and very old-fashioned people, with qualities that I both loved and disliked. I liked the fishermen, I liked the cafés, and I liked the folk dancing in the town hall. I adored the small rowboat that I commissioned a young man to build for me, but as a single woman, I was very lonely. Someone said to me, "If you don't shack up by September in Woods Hole, you will be alone till June." As far as I could tell, this was true. At Woods Hole, I would always miss the sense of belonging I experienced with my cohort of fellow Scripps graduate students.

During Woods Hole dinners and receptions, the men (the scientists) would gravitate to one room, and their wives would gravitate to another. I didn't know where I should go. I had nothing in common with the wives, nothing to talk about. If I went to talk with the men, my colleagues, I would feel the eyes of the wives piercing my back. Ballard might have eased the social tension, but he was largely absent because he rarely socialized with the other scientists.

It seemed to me that the animosity I felt at Woods Hole emanated not from the men with whom I worked, but from their women.

Today, I am convinced that in the past, women were actively involved in holding other women back, even as equal rights were gradually and inexorably taking hold. Many women felt terribly vulnerable since they lacked occupations of their own, and they were very fearful of losing their husbands. In the 1970s, when I would return home from a trip at sea, I felt the open contempt of my colleagues' wives. As a member of the expedition team, I was perceived as a threat. This primal feeling, the jealousy and alienation I sensed from other women, created a sadness within me. My friends were mostly men in those years.

The relationship between the scientists at Scripps and those at Woods Hole was characterized by both strong rivalry and persistent collaboration. Earlier, the Scripps Deep-Tow Group had established connections with Charlie Hollister's Woods Hole group and other individuals working on mid-ocean ridge tectonics. However, there was competition between Ballard's camp at Woods Hole and Spiess's camp at Scripps, and this put me somewhere in the middle. Ballard and Spiess wanted to be central players in the search for the *Titanic,* and since I worked for both of them, I found myself in a rather sensitive position.

In addition to this competition, yet another rivalry erupted, and again, I was in the middle. The Scripps Deep-Tow Group, rife as it often was with internal backbiting, nevertheless had a certain air of camaraderie and loyalty. There was a rule that all original Deep-Tow records must be stored at Scripps; only copies were allowed to be given to investigators from other institutions.

For many years, Scripps scientist Peter Lonsdale had worked closely with Charlie Hollister of Woods Hole, but by the late 1970s, animosity had built up between them. At the end of a cruise in 1978, Hollister evacuated the ship quietly in the dead of night, with all the Deep-Tow records stashed in his suitcase. Lonsdale was furious. Some days later, and much to my surprise, Peter Lonsdale appeared at my office at Woods Hole. Our friendship was an uncomfortable one, made all the more tenuous with the events of the *Pleiades* expedition. Our working relationship was practically nonexistent: I respected him for his brilliant science, but I would never work with him. I couldn't imagine what he wanted now.

Peter outlined a proposition and pleaded with me to help him retrieve the Deep-Tow records from Hollister's office. "Kathy, everyone knows we dislike each other, so *nobody* will ever suspect you if you help me out."

It was hard to resist such a proposal, preposterous as it was. I had my own grievances with Hollister after his actions on *Expedition NATOW;* moreover, I did feel a loyalty toward Scripps. To my own amazement, I could not help but laugh, and I agreed to drive the get-away car to the Hyannis airport. Charlie and Peter really deserved each other. Peter and Kim Klitgord managed to grab the data from Hollister's office and tossed the entire package into the back of my Datsun, which I had parked just outside of the oceanography build-ing. We sped away. Some time afterward, I received the only thank-you letter I would ever receive from Peter Lonsdale: "You must get back to Woods Hole from time to time to pick up your mail, so your chances of receiving this before Christmas must be fair. Mostly, I just wanted to thank you for driving the getaway car. It all seems to have worked out, and I am now sitting on most of the data (by the way)."

Despite the scientific intrigue, I thought I would remain alone at Woods Hole, finding solitude and solace in the sailing and swim-ming. I loved the old Victorian houses with their labyrinths of rooms. I loved the books hidden in the crannies of those houses, the hurricanes that blasted into the coast, the secret swimming coves.

And then, I met and fell in love with the man I thought I would marry. Tom was not an oceanographer but came from a wealthy family who had established a summer home in Woods Hole in the early 1900s. Every year, he hosted a huge Victorian croquet party on the lawn of his family's estate, which sat overlooking Little Harbor. That is where we met, after I smashed his blue ball into the wall of trees bordering the large lawn, down by the Chinese bell and the library boathouse. His grandfather had been ambassador to China, and his father had been a secretary to the president of Czechoslovakia, before the Communists invaded. Through him, I found an escape from the otherwise enveloping world of ocean science. We traveled together, and I met his broad circle of East Coast friends with whom he barely studied when they attended Harvard.

His father and I developed a strong rapport. He was in his eighties, and during most of the year he lived alone in a tiny apartment in Rome, Italy. He returned every summer to Woods Hole, and since we both loved to swim, he and I would take to the sea every morning, very early, and swim from Juniper Point out to a buoy. The summer of 1978 was very beautiful.

The relationship with Tom and his family drew me farther from the everyday routines of Woods Hole. With the exception of Bob Ballard's group, I lacked any substantial contact with the rest of the staff, and this ultimately led to problems. The discomfort I felt among the mostly male scientists kept me from developing plans for future collaborations with them. Because of the independence I had fostered at Scripps, I preferred to work on solitary projects or with investigators from other institutions. I had gone to Kenya with a researcher from Scotland, and I went to sea as often as I could. The Woods Hole Department of Geology and Geophysics did not see much of me, and when I was there, I was very quiet and reserved. I know now that this was a mistake, but at the age of twenty-six, I was disinclined to become politically engaged, and I did not reach out to others, even for reasons of career advancement.

My personal relationship with Tom forced me to reassess my priorities. I weighed the importance of working in oceanography against the desire for a personal life. The men at Woods Hole seemed to have both, and I wanted the same thing. However, balancing the competing demands of my career and my relationship with Tom required more energy than I could muster. Tom disliked my journeys to sea, which lasted for a month or more. For him, my absences were unacceptable, but for me, the expeditions were absolutely essential to my job.

Bigger problems loomed. Bob Ballard was angling for tenure and was having difficulty with the Department of Geology and Geophysics. He opted to move to the Ocean Engineering Department, where he believed he had a better chance at achieving tenure. It seemed to me that a combination of jealousy and conservatism had turned the Department of Geology and Geophysics against Bob, but whatever the reason, when he left to join Ocean Engineering, I was left behind, stranded. Although I was Bob's postdoctoral researcher, there was no space for me in Ocean Engineering.

Soon thereafter I was summoned to the office of the chairman of the Department of Geology and Geophysics. He told me, "Kathy, there is no niche for you in our department."

I was in shock. I had several research grants: one to dive to the Galápagos Spreading Center, one for research at the Tamayo Fracture Zone in the mouth of the Gulf of California, another at the

Oceanographer Fracture Zone, astride the Mid-Atlantic Ridge, as well as a project to map the geothermal regions of East African Rift Valley. I had no lack of funding. I believe it was entirely personal that I was asked to leave Woods Hole.

I was in a panic. The loss of my position would mean a separation from Tom. I did have other job possibilities: at Harvard, for much less money than I was making at Woods Hole; at the University of Miami, which was too far away; and at Rutgers University. Then I remembered the words of an old acquaintance from Scripps, Roger Anderson, who was now at Lamont-Doherty Geological Observatory of Columbia University in New York: "If you ever need a job, give me a call."

That's what I did. When I explained my situation, Roger paused and said, "I'll call you back in fifteen minutes." A few minutes later, Roger called back with Bill Ryan, who offered me a job on the spot. I would work with Ryan's fledgling underwater camera group to map the U.S. East Coast continental margin and the East Pacific Spreading Centers. By the summer of 1979, I had moved to New York but commuted to Woods Hole on the weekends. Tom was really frustrated that I could not stay in Woods Hole, but after all I had been through to get where I was, I couldn't just opt out of oceanography. Commuting five hours between New York and Woods Hole was the best I could manage, and we stayed together for the next two years, until the stress of our physical separation became too great and our relationship ended.

16

ICE PALACE

We're off to the crystalline and cold
where thoughts congeal
in ice wracked
stony silence
only the pure aqua
and splintered radiance remain
drifting
under laid-bare emotions
lying in wait
for life

—KATHLEEN CRANE,
NORWEGIAN-GREENLAND SEA

Of the three great oceanographic institutions in the United States—Scripps, Woods Hole, and Lamont—Lamont fit my personality the best. Curiously, not too many people outside of oceanographic circles had even heard of Lamont. Although Lamont had experienced turbulent periods in the 1950s and 1960s, by the time I arrived in 1979, it was a haven for diverse individuals. The large number of foreigners who came and went through the observatory instantly impressed me, and I rapidly fell in with this lot.

The exposure to international science helped to shift my interests away from the Pacific hydrothermal battles in the United States to research opportunities in other areas of the world. The compounding stresses of my personal life, stemming from the death of my brother and my difficulty maintaining any relationships while living a life of incessant proposal writing under the aegis of the National Science Foundation, reinforced this need for a change. I needed something to take me away from loss and emptiness.

In 1980, I had the opportunity to travel to the far north with the King of Sweden and his royal entourage. The "Ice Palace" beckoned.

The adventure started when Leonard Johnson of the Office of Naval Research (ONR) asked if I would be interested in going to a planning meeting in Copenhagen. The meeting would set the stage for a remarkable voyage into the unknown Arctic, and the King of Sweden was footing the bill. ONR wanted the United States to participate. I hadn't really thought about the Arctic before, but the lure was irresistible. There was one problem, though: I was committed to field research in the wilds of East Africa during the same time, where I would conduct heat-flow measurements in the sweltering Kenyan Rift Valley. However, my years of experience in travel planning told me that I could do both, and I found a way to shift my schedule by a week and fly back to New York from Nairobi via Copenhagen.

After one month buried in Kenyan dust, I put the final touches on the geothermal maps I had constructed and packed my belongings. I left Africa on a velvet black Nairobi night at 2:00 A.M. Below and behind me, the giraffes and Masai roamed across the vast African savanna. Ahead of me lay the world of blue cold.

In Copenhagen, I signed onto the Norwegian Geophysics Team. We would be three: Olav Eldholm from Oslo would be in charge of gravity and magnetic measurements, Eirik Sundvor from Bergen would be in charge of seismicity, and I would be in charge of heat flow. We were to be a part of the larger marine geology team headed by Kurt Bostrom and Jörn Thiede. The team was supplied with its own container, where we would set up shop. The container was actually like a portable cottage, with dark wooden siding, red shutters, and green curtains, and all the comforts of a Norwegian mountain vacation hut. We stored all our equipment, parkas, huge mittens, giant socks, boots, and bottles of 100% pure Norwegian mountain water in that container.

I so adore the warmth of the tropics. How did I convince myself to spend years and years in the Arctic world? The answer is its unique beauty. Vivid greens and blues radiate throughout sharp ice splinters. Delicate pastels glimmer under a midnight sun. Sunrises and sunsets merge together, delicately pulsating. The celestial brilliance of the aurora borealis streams through the ionosphere down to the sparkling ice below. The high Arctic feels pure, limitless, and unknown. I had never realized this until I saw it all for myself. And I was hooked as thoroughly as a fish on a line.

The unexplored and the unexplained draw me. This innate characteristic inspired me to seek the underwater hot springs. Once we located the hot springs and mapped their coordinates, others inevitably followed—the validators, the detail scientists, the statisticians. In time, everything will be carefully analyzed, categorized, and dissected. When this happens, some of the original beauty will vanish, and this saddens me. In the vastness of the uncharted Arctic, there is still newness and there is still beauty.

Many explorers in history—Martin Frobisher, John Davis, Henry Hudson, and William Baffin—braved the extreme conditions of the little ice age (1550–1850 A.D.) in search of the elusive Northwest Passage that was believed to connect the Atlantic and Pacific Oceans. Peter the Great sent his most intrepid voyagers with Vitus Bering on the Great Northern Expedition to Arctic Russia and the Bering Sea (1725–1742 A.D.). John Franklin's expedition (1845–1848) disappeared off the shores of Canada. Fridtjof Nansen and Hjalmar Johansen finally made it back to Norway in 1896 after being separated from their icebound vessel, the *Fram,* for one year. Nils Nordenskjöld

Arctic Region

strove to navigate the Northeast Passage (1879) through regions north of Scandinavia and Russia. In 1909, U.S. Admiral Robert Peary claimed that he had reached the North Pole. In 1917, the new Soviet government developed the shortest water route across the Arctic Ocean north of the Soviet Union. In 1926, the Soviets claimed all land and "unmovable" ice formations in the triangle between the eastern and western most tips of Russia and the geographical North Pole.

In the twentieth century, military and ocean scientists began to overtake the Arctic. While depth charts had not improved much since the days of Nansen, every discovery was a major breakthrough in the understanding of the tectonic-climate history. Starting in 1937, the Soviets established several drift stations to gather Arctic oceanographic and geophysical data. Later, the United States embarked on its own Arctic expeditions with the *Arliss II, T-3,* and *Fram* ice stations.

Scientists discovered that the Arctic Ocean, the smallest ocean, is nearly landlocked. The central polar basin, about twice the size of Alaska, is surrounded by a wide continental shelf, which is interrupted only by a deep ocean trough between Greenland and Spitsbergen (Svalbard), known as the Fram Strait. The shelf from Greenland to Barrow, Alaska, is about 50 miles wide, but in the Chukchi, East Siberian, Laptev, and Kara Seas, the shelf is typically about 400 miles wide.

The Lomonosov Ridge divides the deep Arctic Ocean into the Canada Basin, where the greatest depth is 4,000 meters, and the smaller Eurasian Basin, where the greatest depth is 5,100 meters. The ridge is possibly a fragment of the former Asian continent that splintered off while plate tectonics started shifting the jigsaw pieces in the north hundreds of millions of years ago. The only known volcanically active spreading center that runs through the Arctic is the Nansen-Gakkel Ridge, the northward continuation of the Mid-Atlantic Ridge. It enters the Arctic through the Fram Strait and abuts Russia at the intersection with the mouth of the Lena River.

The dynamics of the Arctic Ocean are known to affect the entire climate of the earth. Large quantities of freshwater flow into the Arctic through rivers and runoff from neighboring continents. The balance of fresh (light) and salt (heavy) water controls the rates and magnitude of sinking surface water down into the deeper parts of the Norwegian-Greenland Sea. This sinking, deepwater process is enhanced when winter approaches, ice forms on the sea surface, and a heavy brine layer collects below. The heavy brine sinks rapidly. To replace the deepwater formation, shallow tropical water is drawn up to the Arctic from the equatorial Atlantic Ocean forming the Gulf Stream. This vast warm ocean current bathes Europe and keeps its climate temperate. Should anything happen to the freshwater-saltwater balance, deepwater could stop forming, forcing a shutdown in the Gulf Stream. Europe could be plunged into a mini-ice age. Ironically, as the climate changes through global warming, more freshwater may pour off of the melting sea ice and out of rivers in Canada and Russia, feeding the Northwest Passage and the North Atlantic Ocean with masses of buoyant surface water. Buoyant water will not sink. As the earth warms, Europe could cool rapidly.

In 1980, not much was known about how Arctic processes affect the stability of the global climate. However, the need to understand these interactions became apparent after the world was frightened by a rumor leaked to the West that the Soviet Union wanted to reverse the flow of its rivers, from the Arctic north to irrigate the deserts of Kazakhstan in the south. Oceanographers mobilized. International governments and Soviet scientists alike protested this idea. If great quantities of Soviet river water disappeared from the Arctic, the protecting lid of springtime freshwater would vanish, allowing the warm Atlantic saltwater to rise and melt the Arctic sea ice cover.

Partly in response to this international incident, and wishing to celebrate the centennial anniversary of Sweden's own Arctic expedition led by Nordenskjöld, the King of Sweden and the Swedish Royal Academy of Sciences invited scientists from ten countries to participate in the first truly international expedition to the Arctic. The Soviets, however, did not show up.

The expedition into the Arctic ice, YMER 80, transported me back to the nineteenth century. From the Portuguese Azores, I flew to Spitsbergen (a Norwegian protectorate in the Arctic) to await the ship's arrival in the desolate Isfjord near the Norwegian coal-mining camp of Longyearbyen. It was early August, and a blizzard was raging. Small Spitsbergen reindeer, unfazed by the weather, nosed around our cabin. Snow poured down the chimney's roof hole because the facility owned by the Norwegian Polar Institute had not yet been winterized. I shared the cabin with two Russian glaciologists who had traversed the mountains from the nearby Russian town of Barentsburg to purchase food supplies (Russia has the right to operate coal mines in the region). A Frenchman and an Irishman joined us later, as we strove to keep warm in the cabin's steaming kitchen.

All of us were waiting for the weather to break so that we could make our way out of Spitsbergen. I learned my first Norwegian words in that cabin. All day long, the radio was tuned to the one available station that mostly broadcast the weather report: "Skiftene skydekke, enkelte regnbyge" (shifting clouds and some rain), over and over again.

For three days, we stayed in the cabin, drinking vats of Russian tea and trading stories. Those three days passed slowly. My Azores tan faded, and my skin seemed to thicken against bitter cold, when miraculously, my giant ship, the *YMER*, arrived. It was grand: eight decks high, so broad, so clean, and so royal. I boarded the ship, my suitcase filled with clothing quite different from that taken on any of the other expeditions I had participated in. The Swedes had requested that I bring a ball gown, for dinner with the king, and the Norwegian team had requested that I bring my toughest winter boots, for days we would spend on the ice. In addition, waiting for me on board, and custom-made to fit my body, was the ultimate in Swedish Arctic wear, from long underwear to thick work pants, a complete suit. It was not elegant, but it was colorful and would keep me warm. Geologists wore orange suits, physical oceanographers wore blue, and biologists wore green. I was dressed in orange.

There I was, complete with boots, Arctic gear, and ball gown, all the things a Polar scientist would need on the job. The *YMER* anchored out in the Isfjord, and a boat was sent ashore to pick me up as it off-loaded other scientists. There was a cheering section lining the starboard side of the ship as I climbed on board. I was the last woman to arrive, and that was that. The *YMER*'s engines revved, and I was shown to my beautiful cabin, with more living space than my apartment in New York City. An icebreaker, the *YMER* had been moved from the Baltic Sea but had never been in the Arctic before. Soon after leaving port, we entered into the Arctic Sea ice cover, shuddering at each jolting impact, our speed reduced to three knots as we moved erratically to starboard and then to port. Massive ice plates shoved the ship as if it were a subway car. Suddenly, a silken thread of water unzipped before us. Small seals darted in and out, leaving their V-shaped wakes on the unruffled surface. Jumbled ice ridges piled high on top of each other.

*King Carl XVI
Gustaf of
Sweden*

The *YMER* was supremely well organized, doubtless because the King of Sweden was on board and oversaw many of the ship's day-to-day duties. Loudspeakers would crackle as the captain announced the next station, the work detail parties, and the exact order in which equipment would be lowered onto the ice. Aside from the king, the 100 people on board included scientists, crew members, an admiral, and several researchers in the humanities—people from nine countries. Each science team was small, but almost every separate discipline was represented: chemistry, physics, marine geology and geophysics, marine biology, atmospheric science, polar bear studies, continental geology, glaciology, and air-sea-ice dynamics.

In Norse mythology, *YMER* is the father of all giants. True to form, the icebreaker was formidable in size. *YMER* was 100 meters long (the length of a football field), weighed 8,000 tons, had twin propellers fore and aft, and was capable of 22,000 horsepower. She required a crew of sixty, and for this summer-long voyage, she would take on ninety-five scientists at different intervals. Thirty-five of the scientists were non-Swedish, and only two of us were women. We were all involved in basic research: search, find, and record. The expedition took on the tenor of the British Challenger expedition, which had mapped and sampled the world's oceans in 1872. Top priority had been given to six broad research areas:

1. the geologic history of the Arctic Ocean
2. the extent and chronology of ice sheets during the past few million years
3. the chemical and physical oceanography of the Fram Strait and the coast of Franz Josef Land
4. the heat exchange between the Arctic environments and the rest of the world

5. the evolution of the Arctic Ocean flora and fauna
6. the extent of pollution in the Arctic region

The never-ceasing cold, the frozen hands and feet, made working in the Arctic frustrating. Equipment refused to function, batteries died early, and ice floes cleaved wires in half and smashed delicate instruments against the hull of the ship. It was impossible to
tow any instruments behind the ship. We had to be satisfied with
one measurement at a time, and we were lucky if anything
returned on the end of a wire at all.

The days progressed as long cold working hours punctuated by
meals, a few hours of sleep, and the random Ping-Pong match.
Each day was full of wonder, hard work, humor, frustration, and
all the brilliance and foibles contained within a group of humans
isolated anywhere in the world.

By late August, we had been underway for several weeks, working primarily on geology and oceanography. We recorded the daily
successes, failures, and difficulties of conducting research in this
frozen frontier midway between Spitsbergen and Greenland.

August 24: I worked all night, and I'm now tired from hours on
end of coring sediment and measuring heat flow. All night we were
delayed. The chemistry crew requested samples, and this meant
additional time. It was midnight, and we still were not able to take
a core. The geology team retreated to the mess hall where we
devoured enough cheese, caviar, and meat to sustain us against the
cold of the snow and sleet. It was such a disagreeable night, but it
ended well at 6:45 A.M. with the sighting of three polar bears mean-
dering close to our starboard side. The admiral crept up to the
bridge in his purple bathrobe to observe them through his field
glasses. It could have been a scene from Gilbert and Sullivan's
H.M.S. Pinafore.

Tagging polar bears was a major project and one of the king's favorites. In the past, Eskimos and furriers had hunted polar bears using primitive weapons. By the 1950s and 1960s, hunters who collected trophies attacked the bears by boat, airplane, and helicopter. In 1973, Norway, Denmark, the United States, Canada, and the Soviet Union declared a moratorium on commercial polar bear hunting. Finally, in the 1980s, researchers began to map the migratory and population patterns of the circumpolar bears.

The Spitsbergen polar bears roam on the Arctic sea ice. They are often carried hundreds of miles on rapidly moving ice floes. Some have even been found off the central coastline of eastern Greenland. Thor Larson, Erik Born, and Mitch Taylor joined efforts on the *YMER* to tag the polar bears. From *YMER*'s top deck, some seventy-five feet above the ice, we would sight the polar bears. The quiet day-like night would be disrupted then by the start of a helicopter that was on call to assist. Once aloft, the helicopter banked as the tagger aimed a tranquilizer dart at the flank of the bear. Once hit, the bear would be unconscious after about fifteen minutes, and it would be safe to land. The taggers had only one hour to perform all of the tests on the bear. The length and weight of the bear were measured, and its blood was sampled for genetic research. A small tooth was extracted to determine the bear's age, and finally a number was painted on its flank to identify it in the future. The mark would disappear after a few months.

We observed that the number of bears increased toward Svalbard and diminished toward Greenland. In fact, on some islands, bear populations were far greater than had been expected, a positive sign for ecologists, a negative one for geologists assigned to and isolated on those islands for months at a time.

By August 26, we were close to the edge of the ice pack, and we could feel the slight swell from dampened waves underneath us.

The ice floes breathed and sighed with each of our movements. A pearly glow diffused throughout the landscape, punctuated in places by a soft radiant blue that darted from the splintering cracks in the ice. Looking back, the ship lay beached on the ice like a huge giant, and we were Lilliputians in comparison.

Our job was to take numerous measurements of the earth's gravity—on the ice. Pack ice provided a uniquely stable platform for delicate instruments such as gravimeters and strain meters. At each oceanographic station, teams of scientists scattered around the ship either measuring ice movements, sampling water or plankton, measuring gravity and magnetics, or searching for polar bears.

On the following day, the sky was quite light even though our watches indicated that it was close to midnight. We received notice that we had secured a night station, and we had to move quickly to gather geophysical data. We suited up for the subzero excursion, piled on the life vests, and loaded the gravimeter, ice axes, pistols, and knives into the transport bucket. Together with the gear, we climbed into the buckets as everything was lowered from the ship to the ice below. The bridge watch kept an eye trained on the horizon looking for any sign of polar bears. Ice ridges loomed out of the milky sky. The eerie light dripped onto the white blocks lit from underneath by that pervasive blue glow. We crept through the light snow, and the YMER slid away silkily through the jet-black water.

At that moment, we were alone on the ice. The eerie silence enveloped us but for the fumbling with the knobs on the instruments and our steady rhythmic breathing, the two alien sounds that kept us in touch with the human world.

Later on that day, an invitation for dinner with the admiral was delivered to me. Gin and tonic, aquavit, beer, and sherry don't mix well during a night of freezing, hard work on the ice. Yet in this environment we had to blend the social and political protocol with the demands of the scientific mission in order to succeed. The contrast between the genteel environment on board with the treacherous frozen environment outside was profound. Four hours earlier, we had been on the ice, buried in five layers of clothing, measuring gravity and taking ice cores from a small, splintered, newly formed ice floe. The gravity readings were unstable since the ice would surge from the wake of the ship as it sailed away into open water for deep-ocean chemistry experiments. We realized later that we needed to signal the bridge watch so that the *YMER* would return to us on the drifting ice floe. After gesticulating wildly and making radio communication, they returned rapidly. People on the ice have priority. We made it back to the ship just in time for dinner. After I scraped off the grease from the fantail, I climbed out of long underwear, thick socks, three sweaters, parka, mittens, hats, scarves, boots, and life vest and took a quick shower before I climbed into my ball gown for dinner with the admiral.

Every evening, the ship's crew and scientists made their way into one of three different dining halls. The admiral's salon on the third deck was reserved for the elite guests. Works of art hung discreetly on the wood-paneled walls. Lush green plants and soft carpets, the finest crystal decanters sparkling on the sideboard, teak chairs around a matching table, and a sitting room furnished with the best Scandinavian furniture completed the scene. Guests arrived discreetly, wearing suits and ties, with the exception of the two women on board. Each night the chief scientists, artists, writers, and the admiral or the king sat down in elegant and diplomatic style to a mostly pleasant but sometimes tense repast. The expedition's work was put on hold during the admiral's dinner hour. It would have been perceived as an insult if we had requested to be excused because a long-awaited experiment was being conducted during those evening hours.

Most of the scientists ate with the officers or the crew. Our dinner places were assigned weekly. We all ate the same food—only the service was different—but with each lower level of the mess, dinner was exponentially more relaxed. Each evening began with a cocktail hour. The ship's protocol required that we present a colored ticket in order to be served any type of liquor. Some enterprising people collected their bar chits to bribe technicians to work extra hours in exchange for a second glass of wine. The food was excellent because our chefs came from the offices of both the king and the admiral. Our meals often included wheels of Swedish cheeses, fresh bread, caviar, fish, paper-thin pancakes, pea soup, punch, and the most charming of all, the weekly "rotten herring," guaranteed to kill the palate and any other senses.

On Saturday morning, August 29, I was jostled out of my bunk by the loudspeaker: "Wake up, Wake up. It's seven o'clock, time to clean the ship." I was shocked that the scientists were required to swab the decks and wondered what the fifty crew members were doing on board. All scientists had to appear on deck three to collect mops, buckets, and wax. Curiously, this was the only loudspeaker announcement made in English during an entire week. For three hours, and according to Swedish naval tradition, we were reduced to cleaning . I remember asking Eirik Sundvor, my geophysics partner, how to wax a floor—because I had no idea. His answer: "No problem. I watch the cleaning ladies outside my office daily. I think we pour the water out like this." Buckets flew and water fell, cascading down the ten flights of stairs. Shrieks echoed from below. We had made a direct hit onto some poor soul.

By the end of August, the sun had dipped below the horizon for the first time in the summer. It marked the beginning of the polar night. Peter Wadhams and Verne Squire from the British Scott Polar Research Laboratory were working around the clock, mapping and documenting the extent, thickness, and coherence of a tremendous ice eddy, like a slow-motion, semi-frozen whirlpool, centered over the Spitsbergen Fracture Zone. They were trying to understand the dynamics of how the large eddies formed and how wave action stimulated strain and the large breakup of the ice pack. We had just finished mapping a transect along the eastern half of the Spitsbergen Transform Fault. Part of it, the Molloy High sits astride the fault, and we were wondering if it might be a large volcano that blocked the trace of the fault. The heat flow data indicated that there was warm water moving upward through the sediments. In this region, heat flow was nearly five times the earth's average. In fact, all the evidence suggested that the ocean floor from Spitsbergen to the middle of the Fram Strait was very hot, whereas the coast of Greenland was cold. The next day we would attempt to break through the thick ice to reach the Nansen-Gakkel Ridge, the active volcanic spreading center in the Arctic.

By midafternoon on August 30, we were caught in old gray-banded ice. Light-blue ponds of water sat atop the floes, and once-liquid rivulets and streams lay frozen under the late summer sky. We were at 82° N off Greenland's coastal Station Nord. A glacier lay sparkling under the low-angled sun. Ahead, the severe and dark cliffs named Esquimonaesset, "the Eskimo's nose," loomed. On top, a glacier rose out of the flannel fog into the rays of the sun and then plunged black and white into the sea. Our Finnish engineers were supposed to leave the ship here, and Valter Schytt, the chief scientist, would visit with the Danish admiral at Station Nord. While the upper command and royalty performed their diplomatic functions, we attempted to construct a camera station on the Greenland Shelf.

Working with deep-sea cameras is a frustrating task. First the strobe breaks, then a relay switch clogs, the batteries wear down, and the data logger jumps into negative time. On this day we were lucky to get the camera back out of the water at all. After one and a half hours on the bottom, ice sheets, like snapping jaws, pressed in on both sides of the ship. As the instrument approached the surface, we could see the strobe light flashing away under the solid sheet of ice, but we could not pull it through the crack of black water between the ice sheet and the ship. The wire was bent taut against the hull, and the pinger was hanging on by only one clamp. Inch by inch, we wrestled the camera from the sharp knife-edged ice. With three hooks and supports, we pried it free. One strobe lay dangling precariously.

Days passed. The YMER continued to crash through the ice, traveling further north. On September 2, at 3 A.M., I awoke to a sharp shout: "Disaster! We've lost everything!"

My first thought was that we were sinking, and I flailed about for my life jacket. But we were not sinking: the entire 12-meter piston core and heat flow apparatus had snapped off from a weak spot in the wire and was resting on the bottom of the ocean at the Arctic Mid-Ocean Ridge. My second thought was, "We might as well be sinking. I think I just lost all my equipment." My third thought was, "My God, I borrowed that instrument from Roger Anderson. What will Lamont do?"

We would be out of commission for a while. To help speed operations, teams of scientists worked the fantail to spool meter after meter of new wire onto the huge reel. Geologists, bird watchers, polar bear taggers, physicists, chemists, and biologists worked night and day through extremely cold weather. Wrapped in thick clothing, others slaved to put new instruments into line, soldering metal, replacing mirrors, fighting an infinite number of calibrations, spending hour after hour making adjustments in a dark room, looking at pale lights running across a small piece of film, cursing and crying in frustration, splicing wire, gluing and taping, scrounging anything that might work. That day and night were not pleasant. At midnight, an iceberg rammed us sideways and smashed a 15-meter dent into the port hull, a massive blow. Loose objects flew in long trajectories across the ship. I woke up smothered under the impact and weight of two huge life jackets.

By September 4, at location 81°31' N, 0° E, the cold was intense. It was −7°C, and all the hoses had frozen, making it nearly impossible to clean off the sediment cores. The chemistry van was filled with more than an inch of slush. Only a sliver of a moon lit our late-night gravity excursion on the ice.

By September 11, the fog had closed in thick and deep. Snow, fog, wind, and ice blended together. The physical oceanographers spent a few agonizing days trying to locate current meters under the newly formed ice sheets. The first method of retrieval was to track the instrument acoustically. After it "talked" successfully to the ship, it was released from the bottom, close to the position of the ship, we hoped. Under normal oceanographic conditions, the instruments were usually spotted as they surfaced. In the Arctic, retrieval is more challenging, and often not successful. Sometimes we had to break through layers of ice. We could see the buoys, but we couldn't get to them. Instead, the brawniest of the crew would be lowered from the ship carrying pick axes, so they could smash through the ice cover until the instruments could be retrieved.

With every day, we moved farther east toward the territorial waters of the Soviet Union. We would sail close to but not travel into Soviet territory. By September 16, we were too close to Novaya Zemlya, the Soviet nuclear weapons northern test site, and Soviet military aircraft flew overhead as a warning. I watched the ocean. It was a wonder how the ice floes flattened out the large Siberian swells.

Crew and scientists retrieve current meter from the ice.

Slush crested on each wave, and a few carousing black seals left silken streaks across the water. The moist chilling air knifed its way directly through our bundled bodies huddled over the gravimeter.

The sharp winds from Siberia and the still whiteness enveloped us, and the ship disappeared, leaving us alone on the ice again. Only the beady black eyes and legs of a curious ivory gull danced across our field of view. By 6 P.M. we encountered complete darkness. Now, we would use lights for all work outside. We were on the same longitude as Istanbul. In two hours we could be at the longitude of Spain.

During the Siberian swells, the giant sediment core broke loose and smashed the supports for the ship's remaining equipment. Meter-wide boxes bounced like Ping-Pong balls. A helicopter rotor was severely damaged, and the electronic readings of the depth recorder jumped up and down with each roll of the ship. Queen

Sylvia's portrait broke loose from the wall in the officers' mess, yet King Carl XVI Gustaf remained staunchly calm and unruffled.

Amidst the wild seas, we attempted more geological and chemical experiments. To our surprise, the core came back without a sample, and all of the heat-flow thermistors strapped to the outside were scraped off of the barrel. This ended the expedition's thermal research program, since all of the spare parts had been lost earlier.

By the middle of September, the temperature had risen

Recovery of hydrographic equipment, Arctic Ocean, 1980.

to –4°C, and long-period swells coursing from the Soviet Union carried huge ice blocks on their backsides. The meteorologists were detecting more particulate matter in their samples, more signs of massive air pollution coming from the northern Soviet cities. A storm system centered at Hopen, between Svalbard and Tromsø, left us with a 60 percent chance that we would hit worse weather on the way back to Norway. As long as we were in the ice and in the lee of Franz Josef Land, we would be safe.

A week later, the temperature had risen to 0°C. We were going home. By the end of *Expedition YMER,* I had gained an understanding not only of a new kind of science but of diplomacy, and how to drink aquavit and still work all night long.

One of the greatest obstacles in understanding the Arctic is not the physical difficulty it takes to get there but the political problems of working in a region dominated by the militaries of the Cold War. Before the Cold War ended, Western scientists knew little of the wealth of information that Soviet oceanographers collected about the Arctic environment. Western scientists had one view of the Arctic and its ecosystems, and Soviet scientists had a completely different view, and in the 1980s it did not look as if both views would ever be merged. Our efforts on *Expedition YMER* did nothing to help the political stalemate between East and West.

In 1981, the Royal Norwegian Society of Technology and Natural Sciences extended an offer to me to live and work in Oslo. I accepted but went for only six months, so that I would not have to relinquish my position at Lamont. In fact, Oslo seemed to be Lamont's far northern outpost. Lamont's director, Manik Talwani, had worked there, and Olav Eldholm, a Norwegian professor, had studied at Lamont. John LaBrecque, another Lamont scientist, also had offices in Norway from time to time. During my stay there, I learned how to converse in Norwegian, as I assembled our data from the *YMER* expedition.

Today, more than twenty years later, I continue to work with Norwegian colleagues. It is always a great pleasure to travel to Oslo, watch the boats, and visit with friends. After that initial foray in Norway, I switched from purely American-based research projects to mostly European oceanography. It simply was much easier to do this than to fight for U.S. research ship time and to compete for elusive stature among the scientists of the aggressive hydrothermal vent community.

The Arctic remains a virgin territory, and that suits me well. We have only a hint of the wealth that lies within its ocean margins.

Submarines, once exclusively dedicated to military missions, have mapped previously uncharted terrain, and Arctic countries are intent on exploiting its resources. But the biggest questions remain. How can we protect this beautiful, remote, and delicate ecosystem from the devastation by humans? Lodged deep within the Ice Palace lie the keys to Earth's climate, and maybe more important, in its beauty lies the hope of humankind. Without the preservation of our planet's beauty, our species is not worth its weight in salt.

17

COLLAPSE

In early November of 1981, a debilitating neurological illness suddenly struck me. To this day, the exact nature of the illness has not been diagnosed. Many doctors suggested that I suffered from multiple sclerosis. Others said not. Suddenly, I could not walk up to my apartment, which was on the top floor of a huge house located on Tweed Boulevard above Palisades, New York. Nor could I walk up the steps to my office in Lamont Hall. I couldn't feel my feet, my legs, or my hands. Sometimes my vision would blur into a double mess of chaos.

It seemed as though I was just steps away from total paralysis. It terrified me. Outside of the death of my brother, these moments were the most devastating of my life. I could not work. But just when I thought there was no one to care for me, my dearest friends and colleagues came through like heroes.

Emory Kristof wrote to me from Washington: "It's tough enough being a world-class oceanographer without the old bod deciding to cut out. Go well friend and may the luck of the toss be with you this time."

At Lamont, seven friends lined up in seven cars behind my house to move me out of my top-floor apartment and into the house owned by Lynn Sykes. Sykes, a professor at Lamont, was on leave, and my friends Lisa Tauxe and Suzanne O'Connell were living in his house while he was away. They took me in until I was able to secure permanent residence with Juergen Mienert, a visiting scientist from Germany, and his family. The Mienerts lived in a cottage surrounded by lilacs, close to my office on the Lamont grounds, far from the roads and close to the warehouse home of Frank Aikman, another friend, who lived on a vast property owned by an eccentric sculptor.

I tried to think philosophically about my physical dilemma. Sometimes, however, the agony was too much to bear, and I would wake up screaming. I feared that I would be trapped forever in a lifeless body.

I spent months of solitude in physical and mental agony. I felt like I was living in a prison. Recovery often seemed remote, if not impossible, but I tried to do something everyday that gave me hope. I was determined to be creative, to survive, and to not allow the darkness to overwhelm me. Sometimes I didn't know what to do but to cry. Juergen, his wife, Elke, my friends, my sisters, and countless others showed me patience and kindness and, in so many ways, saved my life that year. It turned out that I did have the luck of the toss.

By May of 1982, I started to heal, and my young niece Ariel and I learned to walk together. By June of that year, I took my first walk up the long hill to Lamont with my friend Enrico Bonatti. It was the greatest victory of my life.

18

SAINTS AND NAVIGATORS

I spent the summer of 1982 regaining my strength, and gradually, I felt the urge to return to science. In the two years between the fall of 1982 and the fall of 1984, I participated in seven expeditions in four oceans and two seas, from the Red Sea to the Arctic.

On the first two of these expeditions, our team from Lamont used a side-looking sonar, the SeaMARC I, to map mid-ocean ridges and chart the large-scale spacing of volcanism and hydrothermal fields in the Pacific Ocean. We also attached a thermistor string to the SeaMARC I to take water temperature measurements for hundreds of miles along the length of the spreading centers.

The SeaMARC I was a new acquisition for Lamont. The National Science Foundation and the Office of Naval Research declined to provide the funds to acquire this much-needed side-looking sonar, but this did not deter Bill Ryan, my new employer at Lamont. He cleverly joined forces with Jack Grimm, an oil magnate who was interested in searching for both the *Titanic* and the "lost city" of Atlantis. Jack had the money to build the equipment, but he needed a team to operate it. The deal between Jack Grimm and Bill Ryan was rather complex. When Jack wanted to search for the *Titanic* and Atlantis, our team had to follow him out to sea. The rest of the time, the Lamont scientists could operate the SeaMARC I anywhere they wished.

Kathy and the SeaMARC I computer rack, 1984. PHOTO: *Anita Brosius-Scott, Lamont-Doherty Earth Observatory*

Because our Lamont SeaMARC I team was superb, we were able to construct a very creative track, the first search along an entire length of spreading center, and we discovered some wonderful characteristics about the spacing of volcanism along the mid-ocean ridges. Thermal data suggested that, indeed, the high points on the ridges were volcanically active. It had been sheer serendipity that our first research on the Galápagos Spreading Center had been focused on a volcanically active high point. The site was almost exactly 200 nautical miles from the nearest Galápagos Island, in international waters.

The data from these expeditions confirmed that the spacing of underwater volcanoes along the length of the spreading centers in the Eastern Pacific Ocean is nearly harmonic; that is, volcanoes erupt like beads on a necklace out of the mid-ocean ridges and are regularly spaced, with tens of kilometers to hundreds of kilometers between them, depending on the mid-ocean ridge. What was the cause for the harmonic spacing of volcanic activity? Was this pattern repeated at other mid-ocean ridges of slower spreading rates? If so, we may have discovered the key to how heat dissipates from the earth's mantle to the seafloor and the ocean above. The prospect thrilled me. Deciphering the geographical patterns of volcanic activity in the deep ocean would allow us to predict the location of sites of hydrothermal activity.

Enrico Bonatti, on the Red Sea

While at Woods Hole, I had worked in Kenya on the East African Rift (thought to be a terrestrial version of a mid-ocean ridge). Our exploration by aircraft revealed that volcanoes and hydrothermal activity in that location were also harmonically spaced. This analogy using a laboratory model (proposed by the British physicists Lord Raleigh and Sir Geoffrey Ingram Taylor) was intriguing, and I discussed the implications with other Woods Hole scientists, but it generated little interest and I delayed submitting a scientific manuscript on the subject to a geophysical journal. I should not have.

When I reached Lamont, I finally decided to submit the paper. I used the East Pacific Rise and the Juan de Fuca Ridge as examples; the data we had gathered using the SeaMARC I showed a very clear correlation among topography, temperature, and volcanic activity. Unfortunately, the *Journal of Geophysical Research* rejected the paper.

The rejection was demoralizing, but my friend Enrico Bonatti, a Lamont senior scientist and former director of an institute in Bologna, Italy, convinced me to collect more data as additional proof. We would participate in an Italian expedition along the slow-spreading, comparatively young Red Sea Rift in the winter of 1983.

Italian oceanographic expeditions are different from American expeditions in many ways. The Italians do not work at night, so only the Americans on board ran the echo sounder during the dark hours. An espresso machine sat strategically next to our equipment, and by dawn, we were completely wired on that black, viscous coffee. There were no night rations on board, and we were ravenous by daybreak, at which point only a small Italian breakfast of white bread and even more coffee was served.

Lunch, on the other hand, was a formal multicourse affair. Every afternoon, we were summoned to the mess hall and seated at the table. There was dead silence until the captain started to hum a tune or sing a stanza from an Italian opera. He would pause, point a finger at an unsuspecting soul, and demand, "Name this aria."

We were not allowed to eat until his question was answered correctly. When Americans were asked, a very long time would pass before we could put a piece of veal parmigiana or spaghetti into our mouths, because none of us knew the answers. By the time the Italians grew weary and blurted out the response, I was cranky from too much caffeine and unbelievably hungry. These lunches often went on for two hours, after which we were so full that we could hardly work.

The ship's crew was huge. Whereas a crew of nine to twelve people would run a Norwegian ship of equal size, a crew of about forty ran the Italian ship, the *Bannock*. In addition, the Italians never seemed to organize watch schedules, and usually hoards of people were present at every launch and retrieval. No one wore a life jacket. Sailors would leap over the railing, trusting that a friend would grab their ankles while they cut away viciously at a snarl or even a wire that transmitted data. I lost two thermistor cables this way.

Italy is known as the land of saints and navigators, but the Red Sea presented different challenges to the *Bannock's* crew. Once, when the ship began to roll slightly, the captain decided it was too much of a disturbance, and he paced the deck declaring that all work would stop and that it was time to go fishing instead. Much to my amazement, we took the ship to the lee side of the nearest island, Zabargad, tied it up to the coral reef that surrounded the island, and waited out the slight breeze. It turned out that a team of geologists on the ship hoped to sample very rare olivine crystals from the island. Zabargad is the Arabic word for the mineral olivine, and during the time of the pharaohs, olivine was mined for its jewel-like quality.

The island sits adjacent to a major fracture zone that is a continuation of a tectonic fault running through Egypt and Arabia. The ancient Egyptian city of Luxor sits on this fault and was destroyed several times by gigantic earthquakes thousands of years ago. The island of Zabargad was thought to be a slice of the earth's mantle thrust up by those events, and the Italian geologists wanted to test this hypothesis. The laboratory analyses of the olivine crystal chemistry could provide supporting evidence.

The geologists set up quarters in a deserted concrete bunker built by the Egyptians during the 1973 Israeli-Egyptian war. In fact, the whole island was deserted. Enrico and I wanted to join the team for a few days, and we decided to go ashore. We took a small boat, the *Zodiac,* to the shallowest part of the coral reef and then *walked* across the remaining reef, still very far from shore, pulling the *Zodiac* behind us through the shallow water. Only when we were much closer to the beach could we reboard the *Zodiac* and pull up onto the sand. Zabargad is visited regularly by Arab pilgrim-sailors. They come by boat to honor three holy men who died on the beach a few years before our visit. The sailors had built a sepulchre of coral blocks over the graves of the holy men, and we camped there, the one place protected from the elements. It was eerie sleeping over the bones of three holy men.

On the beach, huge sea turtles came ashore during the full moon. A few turtle carcasses lay strewn about, and their beautiful, empty, amber shells, the size of porcelain washbasins, glistened in the moonlight.

Enrico and I circumnavigated the island. We climbed along narrow talus cliffs to the aragonite peaks above sharks circling in the water below and traversed a long expanse of beach littered with detritus from the world's shipping fleets. Light bulbs and clots of tar, castoffs from ships passing through the distant Suez Canal, bobbed along in the coastal water filled with the red primordial-like ooze of unicellular animals that clumped together, forming carpets that ebbed and flowed along the shore. Finally, we arrived at the geologists' bunker on the opposite side of the island.

Earlier, when we had off-loaded the three geologists, the ship's Italian chefs had filled their small boat with barrels of red wine, loaves of white bread, pasta, cheese, and tomatoes. By the time Enrico and I reached them, most of the bread had turned into a stone-like substance, and we had to slice it using a wide-toothed saw. The geologists were supplementing their diet with fresh fish that they had speared in the sea. Every night, dinner duty switched to another person. Enrico and I were assigned the task together, and we tried to be creative with some cuisine that differed from the

Italian standard. The meal was so unpopular that we were voted forever off cooking detail. I remember Enrico, himself a native Italian, muttering that night, "Boring Italians—they know how to eat only one kind of food."

The final insult occurred when I poured the wine, which was rapidly turning to vinegar, from the wrong side. One of the geologists commented, "If we were in Sicily, they would cut off your hand for that."

We may not have won over our colleagues with our culinary ability, but we did manage to write two articles based on our new data: one for *Scientific American* titled "The Geology of Oceanic Transform Faults," and one for the *Journal of the Geological Society of London.*

After this expedition, with much more data in hand, I decided to submit my rejected paper, "The Spacing of Rift Axis Highs: Dependence upon Diapiric Processes in the Underlying Aesthenosphere?" to the journal *Earth and Planetary Science Letters.* It was accepted and published in 1985, but this occurred a few months *after* a very similar paper written by Woods Hole scientists was published. Being second to publish in science is tantamount to not publishing at all. It was a lesson learned the hard way.

19

TITANIC *MADNESS*

Location: North Atlantic Ocean, southwestern
   terminus of the "Highway to Hell"
Event: *Expedition Titanic,* 1983
Why: Because it is there
Who: Lamont-Doherty Geological Observatory,
   Bill Ryan and Jack Grimm
Money: Too much
How Deep: Too deep

—KATHLEEN CRANE

It seemed like everyone in oceanography wanted to find the *Titanic* except me. Even when I was a student at Scripps, the next big event after the discovery of deep-sea hot springs (or hot vents, as they were called later) was supposed to be the search for and location of the *Titanic.* Many of the same people were involved: Fred Spiess from Scripps, Bob Ballard from Woods Hole, Emory Kristof from *National Geographic,* Ralph White, an underwater cameraman from Los Angeles (who later worked for James Cameron, the filmmaker), and Bill Ryan from Lamont. I had worked for or with all of these people at one time or another. Their motives were all somewhat different.

Fred Spiess was known for building and engineering equipment designed to search for and locate objects on the seafloor. His first big test was the expedition to locate the *Thresher,* a sunken submarine. To Spiess, the *Titanic* was another mysterious object hidden under the waves. Because it was there, it should be found.

Bob Ballard had acquired considerable fame with the Galápagos vent discoveries, and he was angling for more challenges, more rewards. He had worked for many years with Emory Kristof, the *National Geographic* photographer, and together they were seeking funding from the Walt Disney Company to search for the *Titanic*. However, Disney had already approached Spiess and Scripps, which was just down the highway from Hollywood.

Scripps was not unfamiliar with the film industry. In 1976, Paramount Pictures had leased one of the institute's biggest ships, the *Melville,* for use in the remake of *King Kong.* For years afterward, the movie name for the ship, *Petrox Explorer,* showed through the blue paint flaking off the hull of the *Melville.* Furthermore, the studio had established a fellowship for graduate students at Scripps. The arrival of this ship inevitably produced a crowd of *King Kong* groupies in various ports around the world. I think that the Mexicans were the most devoted fans; they came out in throngs to get a glimpse of the famous vessel. Jessica Lange got her big break in this film, and it is always funny watching her love scenes, which take place in the very same cabin in which I slept.

In addition to Disney, Clive Cussler, the author of *Raise the Titanic,* was enlisted to help raise the large amount of money needed for the many trips to the *Titanic* site. Numerous other plans were set in motion to secure the funding needed for such a difficult undertaking.

Actually, locating the *Titanic* was expected to be relatively easy. After all, the *Titanic* was an enormous ship, much bigger than a tiny hot spring on the Galápagos Spreading Center. Certainly it could be found with sonar. It was anticipated that it would be found within months, maybe a year. Although the location of the *Titanic* was expected to be easy, the rewards would be immense

compared to the amount of money invested. And that was one reason for the eruption of conflict between the primary investigating
institutions.

During the late 1970s, the Ballard and Spiess camps had grown
more and more distant from one another. It was evident that they
could not work together. The events leading to the 1979 discovery
of the East Pacific Rise "black smokers" (superheated vented water
colored "black" by metals leached out of the volcanic rocks below)
proved that. Their styles were just too different. They were like oil
and water.

While Ballard, using his remarkable marketing skills, was busy
securing funding from the U.S. Navy and from IFREMER, the
French government oceanographic agency, Fred Spiess joined
forces with Bill Ryan, who had already promised Jack Grimm to
search for the *Titanic* in return for the use of the SeaMARC I sonar.
In 1977, I had gone to Woods Hole to work for Bob Ballard. Then,
in 1979, after Ballard had switched departments, leaving me
stranded, Bill Ryan had rescued me. I was now in his camp. When
he asked me to participate with him on a *Titanic* search mission, I
agreed.

I had absolutely nothing to do with the planning of the *Titanic*
expedition, and it appeared to be an easy assignment. However,
given the nature of the North Atlantic, there was always concern
about the weather. Once I arrived at Lamont, I kept getting phone
calls from Bob Ballard. I had access to every camp, so I was a good
source of information.

The quest for the *Titanic* had pulled Ballard and Kristof to sea in
1977 on the ill-fated *Alcoa Seaprobe,* a drilling ship belonging to the
Alcoa Aluminum Company. It was built of aluminum, and like a
giant soda can, it blew away from the survey site every time a little

Lamont's Titanic *operations,*
SeaMARC I rescue, 1983
PHOTO: *Anita Brosius-Scott,*
Lamont-Doherty

Women's work. Kim Kastens and the author hauling cable off the fantail,
Expedition Titanic, 1983. Bill Ryan supervises in the background. PHOTO:
Anita Brosius-Scott, Lamont-Doherty

wind arose. Ballard preferred to use deep-sea cameras (as he had at the Galápagos), and he attached them to the end of a long series of pipes that were lowered through a well in the ship's hull. One stormy night, the pipe collapsed, leaving the parts like a pile of Tinkertoys on the ocean floor of the North Atlantic. This was a major setback. How would they ever recoup the loss of photographic equipment? Ballard could do little but sit and watch as others took up the search.

Initially, Spiess and Ryan intended to use the Deep-Tow to locate the *Titanic*. An attempt was made by them in 1981, but it was unsuccessful. The weather blew them away, and they had to return home. By 1983, Bill Ryan was ready to try again. Like Spiess, he would use side-looking sonar, but this time it would be the newly acquired SeaMARC I. We had just tested it in the ocean off the Pacific Northwest coast, and now it was time to repay Jack Grimm. Instead of acting as chief scientist, I took the position as watch leader on this expedition, and to my surprise, I found that it was difficult to carry out the survey. I felt growing admiration for Bill Ryan's patience in the face of the overwhelming problems associated with the *Titanic* search. Lamont scientists were skeptical of the search because it was not traditional science. Although the search was not considered to be purely scientific, it did require oceanographic knowledge and skill. Survival in oceanography required a lot of innovation, in this case, Jack Grimm's SeaMARC I.

Although we went out to sea in summer, it was cold and rainy. Storms blew in and out rapidly, and we were besieged with high waves and seasickness. Jack Grimm was frantic for us to succeed, and he had a right to be, given his investment. Monitoring a decent watch was difficult. Normally, we simply would map the seafloor, but in this case, we had to be on the constant lookout for something that looked like it did not belong on the seafloor. The problem was that a huge underwater canyon lined with gigantic rubble coursed through the search area. Some objects were the size of the *Titanic.* Jack wanted to go back to a site where the Lamont group had previously photographed an unusual object on the seafloor; he thought it might be the propeller of the missing ship. However, when we surveyed the area, we could "see" nothing with the sonar.

Sleep offered no respite. My dreams were filled with the screams of the *Titanic* victims, and it unnerved me to be at the exact spot where such a terrible tragedy had occurred. It made me ill to think about it.

Soon a huge storm blew in, and we had to retreat. Although we had not found the *Titanic,* we had created a map of where the *Titanic* was *not.* We had covered 90 percent of the survey area without success. With only 10 percent remaining to be explored, Bob Ballard, Emory Kristof, and the IFREMER team led by Jean-Louis Michel went back to sea, and they took the Lamont map with Bill Ryan's blessing. The French ship *Le Suroit* spent ten days towing her side-looking sonar, the SAR, yet nothing was discovered. Then the Woods Hole ship, *Knorr,* took over. During that final episode, Ballard, Kristof, and Jean-Louis Michel found the *Titanic* on Michel's watch. It was directly underneath one of our SeaMARC I tracks, but it was at the very edge of our survey, in a place not yet covered by the sonar. Bob Ballard would get the glory. Bill Ryan, Fred Spiess, Jack Grimm, Jean-Louis Michel, and Emory Kristof deserved their share, too. The rest is history.

20

A SCIENTIST
IN FILM SCHOOL

Nothing that is so is so.
—WILLIAM SHAKESPEARE

It was time for a change. The seven expeditions I had participated in between 1982 and 1984 were over. I was losing patience with the way in which U.S. science research was conducted, requiring scientists to devote every aspect of their lives to research, and with the Cold War–style of funding, which forced scientific research to become overly competitive and to focus on individuals rather than cooperation. My increasing frustration with trying to do creative scientific work was reaching its limit. In fact, I had been contemplating a move from science to art for several years. Science is one way to look at the world. It represents the rational side of human existence, whereas art represents the contemplative or emotional side. My experience told me that emotions were a far stronger force in life than rational thought, and I felt drawn to portray these ideas through the medium of film. But first, I would have to learn something about this highly complicated art.

As part of the application procedure to the California Institute for the Arts, I was required to make and submit a film for examination by the admission committee. I decided to use the properties of physics and metaphysics as themes. The movie focused on the difficulties of belonging to a group, to a generation, to an occupation, and to a family. Many of my friends were pressed into acting. Enrico Bonatti, who played my main character, portrayed wistful apprehension about taking part in a marriage while remaining an outsider, an onlooker, to this most intimate of ceremonies. My goal was to put these feelings onto film. We set up stage in a landscape fashioned from a sculptor's estate adjacent to the Lamont grounds. The land was littered by giant spheres that we arranged with the sculptor's permission. They were scattered across a meadow and among trees, like groups of gossiping people or isolated individuals. I directed my friends to wander among the spheres resting here and there, just gazing into the distance or looking very closely at the hands of the couple being married. In hindsight, I can see that the subject matter was strange, but the resulting film was good enough to get me admitted into the California Institute for the Arts graduate school of film.

The world of graduate work in art was completely different from the one I was used to in science. The first day I entered Cal Arts, a student had covered the entryway with a pile of cow manure. In fact, he was using the manure to build a statue of the muses.

I remember only three of the professors and almost none of the students. Of the three professors, one had won an Academy Award for a documentary film, one was visiting from Poland, and one was a woman from NYU who made films about angst. I had enough angst in my life, but I was interested in documentary filmmaking, and Cal Arts, a fiercely independent film center despite its proximity to the mainstream Hollywood industry, was the place to pursue this. I learned all about the science of television and film production, the art of imagery and color, and the influence of sound on the senses.

The job of the student was to watch films. Everyday I watched films, each film four times. The first time, we watched a film with all of its intertwined components: light and image, the spoken soundtrack, and the music. The second time, we watched it without any sound. The third time, we watched it with only the image and music. The fourth time, we watched it with only the image.

What a new way to view the world! Actually, there were some similarities with oceanography, where one has to construct the shape of things by using only sound. Maybe my career in science had prepared me for this excursion in art more than I had realized.

The days and months went by. We saw and made film after film. As I socialized with my new colleagues, I learned that most of them knew very little about science or the natural world. I suppose this should not have been surprising. In America, all students who become scientists must study the humanities, but the reverse is not true. When I thought about this, I recalled the many scientists I knew who were familiar with music and literature. With this realization, I became convinced of the need for scientists to become involved in a reverse communication process. To this end, I decided that I should produce documentaries about the earth.

I wondered whether I would ever return to pure science, pure research. One science colleague wrote to counsel me:

> The exhilaration so obvious in your prose tells me how liberating Cal Arts is for you. You must be discovering many things, and I can imagine how drab the world of science seems in comparison. You really are an artist, and so few of your (scientific) colleagues share your sensibilities. You want to create—really form something, out of nothing human—and you get stuck in a system that accepts (and funds) only incremental creation—a society of bricklayers in a world that needs cathedrals. But let me argue a different point—

Don't leave your science behind.

Point one: You are a very talented scientist with the proper eye for scientific problems. You can be successful in that part of science where most of our esteemed companions are quickly lost— creation. Point two: Science has enough meat to hold your interest for a very long time. Is that true of other modes of existence? Forgive me for getting preachy on you, but you know what happens when someone goes to industry: "She'll never return; no one is going to take a factor-of-two cut in pay." You're in the same situation, only the currency isn't money. I, too, am beset with needs I find hard to satisfy . . . the price to be paid by those who step out of line. I wish you were back here.

By the end of the first semester, I had won awards in film school as best editor and most visionary. But the most important part of my move, my personal agenda to obtain freedom through art, collapsed. I realized that with my neurological disease and no one to depend on, I had to have a job that was stable and reliable. Though I was well enough now, I could relapse at any time. I was forced to make a choice. Either I would try to take this film experience back with me to science or I had to give up science altogether. My anxiety about unemployment was too great.

With sadness, I moved back to New York to pick up my old life. Bill Ryan had retained my position at Lamont, and I accepted a professorship at Hunter College, an economic safety net against the fear of another bout of illness. In 1985 it was clear that I would return to science, but this time through the back door.

21

FASHIONS,
FRENCH, AND FULBRIGHTS

I will adopt an attitude of passive decoration.
—DOROTHY SAYERS

The world changes when we least expect it. It was the middle of the narcissistic 1980s, and I was back in New York. Teaching at Hunter College was leaving me breathless and very, very tired. I organized courses ranging from Introduction to Geology to Geophysics. There were hundreds of students from every background imaginable in my day and night classes. A very large majority of these students knew nothing about America outside of the boundaries of New York City. It was a humbling experience.

My first mistake was to assume that these college students knew the names of the continents and oceans. They did not. My second mistake was to assume that my students knew the principles of science. They did not. I grew exhausted trying to imagine how they could not know these things.

My exhaustion led to two things. The first was that I fled weekly to my office at Lamont for respite from these inner-city stresses; the second was that these students inspired me to develop a program of science documentary videos. I wanted to provide them with images of the beautiful world outside their small sphere of experience, to expand their horizons.

By the spring of my first year at Hunter College, I had created the rudiments of a video about the oceans, the heat from inside the earth, and its role in shaping human history. I called this video *Heat*. In 1987 I applied for and received seed money from the Charles Lindbergh Foundation. Hunter College supplemented the project with additional money, and slowly, more and more companies, interested in advertising to my students, offered funds for travel and for room and board during filming. I asked a friend of a Scripps friend to help with the production. He was a fashion photographer in New York, and he traveled among the rich and famous, stars and starlets, fashion models, Prince Albert of Monaco, the Kennedys, and young European aristocrats (whose titles were in name only). This was *not* a world I was trained to handle.

I could handle a winch at sea, but I could not handle a sleek dress. I could handle the 4–8 watch on a ship, but I could not wear high heels. I could handle days and even months slogging through mud and navigating over rough terrain in Africa, but to mingle in the world of fashion was daunting.

It was such a bizarre world to me that I was shocked when my new friend, Doug, asked me out. Our first date was on a Wednesday night, and he took me to a party at the apartment of the former Miss Hawaii. The party was on Manhattan's Upper East Side and didn't even begin until 11 P.M. This was usually way past my bedtime because I had to rise early for morning classes at Hunter College. I went to it anyway. I needed a new life, and this man was unlike any other I had been with. Given my earlier trials with relationships, I appreciated this fellow with a stick-to-it and stick-to-you quality.

So, for the first six months of what would be ten years together, we lived in two worlds. We went to parties with the glitterati at midnight, and I arose at 6 A.M. The circles under my eyes started to grow deep and dark. I realized that in much of the world, it is impossible to meet people with very different careers simply because they sleep during different hours. You could coexist in the same city, on the same block, in the same apartment building, but the 10–22 (10 A.M. to 10 P.M.) watch never would meet the 22–10 (10 P.M. to 10 A.M.) watch.

We spent a lot of time living illegally, behind stacks of cardboard boxes in his Chelsea loft. Sometimes I would not be invited to parties because I would nod off at 2 A.M. Other times, I would attend and would be heartened when a runway model sought me out. My career in oceanography was very exotic and alluring to these people, and I found that I began to enjoy their company. Having a boyfriend who was a fashion photographer rather changed how I viewed myself and, in some cases, how my colleagues viewed me. He replaced my whole wardrobe and put me through a crash course in walking in high heels—for use on the playing fields of Chelsea and Lower East Side parties. Still, I always had a nagging fear that people would find out that I was not one of them.

I remember one party in particular that took place in a posh restaurant. A woman from the famous Italian Borghese family and her boyfriend, the Prince of Yugoslavia (I didn't even know that Yugoslavia had ever had a prince), were seated next to me. I was impressed when Ms. Borghese and Mr. "Prince" ordered the most expensive meals in the house, finished them, gently wiped off their mouths, got up and walked away, with nary a word about paying the bill. What arrogance these people had in New York at that time. This is not to say that all Europeans behaved so brazenly, but this elite group of people was brash. Many had been hounded out of their native countries and had retreated to the lucrative American shores, where people were impressed by old-royalty titles.

I was not completely ignorant of life in the upper echelon. My former boyfriend in Woods Hole was from an upper-class family in New England. His family's house contained gifts from presidents and prime ministers. When we were invited to dinner there, the large dinner table was often graced by luminaries such as Thomas Lamont (the son of the woman who had donated the land and original buildings comprising Lamont-Doherty Earth Observatory to Columbia University), psychiatrist Kurt Adler, or the head of the AFL-CIO, among others.

Moreover, a close friend of mine was a member of the French aristocracy. We had met in the mid-1980s before I went to Paris to carry out research funded by a Fulbright grant. I had taken up residence in a small apartment near the Place de Beauborg, and it was very difficult to meet anyone in Paris outside of the family of my French friend. I have to admit that I did not have fond feelings for the world of the aristocracy at the time, and I was very nervous about visiting my friend's family mansion on the Mediterranean coast for a week's vacation. Of course I went, but I was trembling.

Scarlet bougainvillea cascaded over the gently arched walls of the villa. Below lay the azure sea, ebbing and flowing against the rocky limestone cliffs that lined the shoreline. During my visit, I stayed in an isolated room that had a beautiful panorama of the sea and the surrounding olive and pine trees, but my quarters were connected with the rest of the estate only by an underground tunnel. I was instructed to be dressed and ready for drinks and dinner on the terrace at 8 P.M. At 7:55, a maid appeared at my door and led me through the tunnel to the terrace. My memory may be foggy, but this is what comes to mind: servants; my friend, elegant in his black suit; serious drinks, of which you weren't supposed to have too many; Provençal cuisine; and shooting stars, one after the other. They rained down over the velvet sea, leaving golden streaks in their wake. I felt uncomfortable—I didn't know what to do or how to act—but I imagined scenes of Grace Kelly and Cary Grant in the film *To Catch a Thief,* and I tried to assume that kind of role. In any event, this family was very patient with my inexperience in the matters of French upper-class etiquette.

Back in New York, Doug and I were continuing to make films for those hundreds of students of mine. We filmed in Iceland, Italy, Africa, France, and Hawaii, among other locations. Hotels and airline companies paid for our expenses, which was a shock to me. I would never have thought to ask a four-star resort for free accommodations. No science program in the world would have been supported this way. Filmmakers, however, often worked in beautiful locations without having to spend a dime. Begging for money was not new to me, but this style of funding certainly was. The obvious reason for this discrepancy is that film acts as a kind of promotional advertising.

Shortly thereafter, Eastern Europe began to collapse, and a new wave of fashion models flooded into New York. At the same time, cosmetic and fashion markets started to blossom in Poland, Russia, and the rest of Eastern Europe. One of Doug's acquaintances, a Russian ballerina who danced with the New York City Ballet, convinced Doug and his colleagues that this was a good opportunity to market cosmetics to the people of Eastern Europe and to bring the previously unavailable fragrances of the Polish perfume industry to the United States. I became the chief scout for sales outlets around the world. Doug and I traveled to Kraków, Poland, to check out factories and import hundreds of cases of rare perfume. All of the perfume was stored in our apartment, which soon began to reek.

During every scientific conference I attended, I would ramble the streets of the city, stopping at stores that marketed cosmetics, selling product as I went. The new economic reality of Eastern Europe presented tremendous opportunities, and we were eager to establish connections with the cosmetics industry there—that is, until my illness struck again. Soon I was retreating once again to the economic security of the world of science.

22

1984

STALIN'S STATUE
You are a requiem for
Borders passing under us.
Thoughts—I take you with me.
I take your sadness,
I take your light,
I take your hollow skins,
Your touch and voice
And laughter.
I take your passion
Soul to soul
And a little of your grayness
Now rests upon my brow
Permanently stamped.
A cold Moscow
Touch-up there
Out here
And all around.
   —KATHLEEN CRANE,
MOSCOW-FRANKFURT, 1984

I t was 1984, the year that George Orwell predicted would bring us a completely totalitarian world. In reality, there was the continuous Cold War between East and West. It was the year that Andrei

Sakharov was imprisoned. It was the year of the boycott of the
Moscow Olympics. It was the year when many American scientists
decided to boycott the International Geophysical Union meeting
held in Moscow. However, I decided to go to the meeting and was
scheduled to deliver a paper about our discoveries in the Norwegian-
Greenland Sea. I thought, too, that it would be better to meet with
the Soviets face to face than to further isolate them from us.

During the decades of the Cold War, our world of science had
been driven by East-West animosity, from the U.S. Navy-funded
oceanographic expeditions to the military-style hostility toward
anything Soviet. Our lives were governed by constant subterfuge
and threats of nuclear war, the spar-and-parry politics that defined
the era. In contrast, my curiosity about the Soviet system, along
with the scientific community's nascent efforts toward an open dia-
logue among all scientists, drew me to the Soviet Union. Nevertheless, I was anxious about going to Moscow. MIT's Tom Jordan
and Marcia McNutt also planned to attend, and without their stal-
wart support, I might have decided against the trip. I set aside my
fear and coordinated with Steve Miller, my colleague from Scripps,
to fly to Frankfurt together, after which we would part ways as we
continued to Moscow: He would board a Lufthansa flight, and I
would board an Aeroflot flight. I was even nervous about the flight,
since I knew no Russian words outside of *da* and *nyet*.

I remember that the Aeroflot plane was different from any I had
flown in before. The seat backs all flipped forward, and the flight
attendants dressed in drab uniforms. Since this was a designated
international flight, we were served oranges, a very rare occur-
rence I learned later. As we flew into Soviet airspace, the amber-
colored sky gradually filled with a creeping darkness. The darkness
also rose from the ground, as we left the glittering cities of Western Europe behind.

We landed at the Moscow airport, and as we filed into the airport, I was directed to the line for foreign visitors where I spotted many of my colleagues from France, Norway, and Germany. We waited for hours. When it was my turn to speak with the customs authorities, I smiled at the young man behind the booth, but he responded with nothing but a glum expression. This was a different world. It seemed both strange and obvious that it was going to be difficult to break through these Cold War attitudes that defined and confined our worlds in science and society.

We filed out, collected our baggage, and were greeted by organizers of the meeting. Translators were available, and excursions were arranged for all the participants. I climbed aboard the bus for the Rossiya Hotel and was amazed at what I saw through my window. Beautiful birch forests gave way to the gray of the Moscow suburbs. Soon huge high-rise buildings appeared, many of them crumbling. Street vendors selling fruits and vegetables from their gardens lined the roads as we entered the center of Moscow and approached the hotel. The Rossiya is one of the largest hotels in Moscow and sits on the corner of Red Square behind St. Basil's cathedral and the Kremlin. There were about fifty sets of doors leading into the Rossiya from all sides, but to my knowledge, only one was open. There were about fifteen counters to process arriving guests, but only one of those was staffed, so we waited for hours more. Steve and I were going to share a room to save money.

Once checked in, we walked down endlessly long corridors toward our room. At each corner sat an old lady who observed the hotel guests, watching for any suspicious activity. Each of these monitors had access to our room keys, which meant there was no security or privacy whatsoever.

The hours in Moscow evolved into days, and Tom and I occupied ourselves by strolling along the Moscow River, talking about philosophy, about Russia, and why we both had decided to participate in a meeting that most Americans chose to boycott. Later, we were invited to an elegant reception. Bottles of wine and vodka were piled high, surrounded by mountains of caviar, cheese, black bread, and *blini*. A Soviet geophysicist came over to speak with Tom, and he asked why so few Americans chose to attend. Tom answered honestly, "They chose not to come because of the Soviet persecution of Sakharov." The Soviet geophysicist answered with a snort, "Sakharov is dead," and then walked away.

Steve had a linguistic advantage over Tom. He could speak a little Russian, and this enabled us to purchase milk and bread every morning at the hotel's corner café. One evening we decided to brave a real Soviet restaurant, after Steve procured instructions from a colleague. We headed out, found the building, located its entrance, and took its elevator up a rickety seven flights. Once on the landing, we turned a corner and walked straight down a long hall to find the restaurant at last. We must have been the first foreigners ever to enter the place.

Someone who looked like a waitress grabbed us by the arm and led us to a table already occupied by two middle-aged men. We sat down, Steve on one side of the table and I on the other. When the men sitting next to us heard us speaking English, their eyes grew wide, and we could see that they were very uncomfortable being seated next to us. I don't remember who broke the ice, but we found a common language in German. Our dinner companions were sailors engaged in a fifteen-year reunion. We soon discovered that we had extensive seafaring travels in common, and then drinks of every sort were brought to our table, followed by jokes, then back-slapping and toasts: "For those who are on the sea!"

One of my duties in Moscow was to take papers from Enrico Bonatti to Gleb Udintsev. Gleb was forbidden to have any contact with foreigners, so I went first to see Leonid Dimitriv at the Vernadsky Institute. His office was filled with flowers and paintings, and it epitomized for me the Russian love for poetry and for the earth. Leonid agreed to help me deliver Enrico's manuscripts, but it would take a little time.

Meanwhile, I presented my prepared speech to the congress and received a rousing ovation. I am not sure why, but I think it was because I was one of the youngest women to have given an oral presentation. One woman approached me later to congratulate me and said admiringly, "Spoken just like a man!"

Elena Lubimova, whom I had met earlier in England, came up to me as well. We had planned to discuss the creation of a book about heat flow in the Arctic Ocean, and this was our chance to talk, but it was very dangerous for Elena. Her son was a dissident, and because of this, her desk had been removed from the Shirshov Institute of Oceanology, and she was forbidden to meet with foreigners. She advised me to not say a word, but to remove my name tag from my dress, and to follow her silently to her apartment. Once there, we entered her rooms, which were filled from floor to ceiling with manuscripts and books. She reached under a counter, pulled out a bottle of champagne, and popped the cork right out the window. "To blasting away these awful borders. To friendship," she toasted. We talked long into the night.

Later during the conference, Elena and her husband invited Tom Jordan, Karl Hintz from Germany, and me to dinner at one of Moscow's best restaurants. It was a place where Russians as well as foreigners could eat. The double standard that pervaded Soviet life at the time bothered me no end. Supposedly all Soviet comrades were treated equally, but obviously some people had more rights than others; elites could walk to the front of a line, enter any restaurant, and rent a spacious apartment. There was no equality at all. And Soviet citizens were discriminated against in their own country. Where foreigners could enter exclusive restaurants, Soviets could not. Where foreigners could stay in the best hotels, Soviets were banned.

The restaurant where Elena and her husband took us was a converted railway station with a huge glass dome. Table after table extended over the vast expanse. A very loud band played in the center. The loud music helped to mask most conversation and created an atmosphere where Soviet citizens could converse free from eavesdropping government officials. When we started to eat, Elena's husband quietly and very formally asked us, "Do you believe in God? Do you think it is okay for a scientist to believe in God?"

We responded that yes, it was okay to believe in God, and the Russians made a series of toasts:

"A toast to God who is here in spirit!"

"A toast for those who are able to leave!"

"A toast to friendship!"

What we took for granted in the West—the right to believe in some kind of religion—was very dangerous in the Soviet Union. Our friends risked telling us that they belonged to a church. I am not much of a churchgoer, but their courage touched my heart.

The next day, in the conference hall, I was opening my briefcase, when a tall man came up to me and asked, "Are you Kathy Crane?"

I answered, "Yes."

"I am Gleb Udintsev," he told me. "Please meet me by the Kremlin at 3 P.M. tomorrow."

What could I do? I had to fulfill his request. I walked along the wet cobblestones near the Kremlin wall, and Gleb arrived promptly. He opened his briefcase, took out a sheaf of papers, and handed them over to me, "Please take these to Enrico. Please get them published."

I stuffed the papers into the mix of papers in my satchel, hoping that when I left Russia, the customs authorities would not discover the breach. Even today, I am disturbed by the role that secrecy and subterfuge played between Russia and the United States. The secrecy was emblematic of the evil machinations conducted throughout the Cold War. The ensuing paranoia fueled scientific research and helped to split our worlds apart.

I thought of Elena, and I wondered, Why did the Soviet Union persecute its top oceanographers? Could it be because they were among the first groups of Soviet scientists to work with foreigners on Western ships? Elena was, in fact, one of the first woman scientists to work with U.S. scientists, on a Scripps ship in the 1960s. At that time, a U.S. oceanographic institution would never have invited a woman to participate on an expedition at sea. However, due to cultural ignorance, the U.S. authorities were oblivious to the fact that only female Russian last names end in "a." When the U.S. Department of State asked the Soviets to recommend two of their top oceanographers to participate on a Scripps expedition, the Soviets responded by sending E. Lubimova and G. Udintsev. Nobody at Scripps knew that Lubimova was a woman, and by the time she had arrived in San Diego, it was impossible to reject her. For many years, there existed so much opposition in the United States against women going to sea that I found it amusing that one of the first woman scientists, if not *the* first, to sail on a Scripps vessel was a Soviet.

On our last night in Moscow, a Soviet geophysicist took us on a tour of the city. He pointed out the places of interest, told us how many of his family had been killed during Stalin's purges, and dropped us off at one of the Red Square restaurants open only to foreign visitors. Marcia McNutt was there, and Tom and I joined her at the table. The liquor flowed that night, and we stumbled onto Red Square just as gigantic tank-like street cleaners were fire-hosing the square. Tom was right in the path of one of these behemoths, which looked as though it had escaped from a science-fiction movie. I screamed at him to run, and we both tried to escape as the tanks closed in on us. Tom stumbled and fell in the process, just as some Soviet policemen surrounded us. We had had too much to drink, and Tom lashed out, "Damn the Soviet Union. Leave me alone!"

It turned out that the police tolerated inebriated individuals. They simply helped Tom to his feet and gently shoved him on his way. Red Square would now be clean.

When we departed the next day, everyone applauded as the Aeroflot plane took off. I breathed a sigh of relief and wrote a poem.

WITH THE
NAVY TO THE NORTH

A wild turkey is loping along the dunes of Cumberland Island, off the coast of Georgia, its three-toed gait keeping it supported along the flying sand. I wonder what it is searching for in this remote place. Faster it goes, up the hill, straining its neck as if a forward thrust would help to propel it upward. This is how I walk sometimes, using a thrusting motion to kick-start my ailing legs up and forward.

In 1989, I was hit again by paralysis, and it was enough of a problem this time to entice me to this warm island. I was on a beach at dawn while a hot morning sun rose and pierced my sweater and stretched my shadow long across the sand and the adjacent grass. It made me think of the light in the far north.

Gulls cried plaintively and the surf crashed. The entire world seemed milky, slowly diffused through a rising band of light. Shrimp boats plied the waters while those at the nearby inn slept. The long porch, green and shuttered, lay open to the cool breeze, while the large, blue-cushioned swing swayed back and forth, back and forth, like water lapping against a boat's hull. Beyond the inn, a kingfisher sang to a crow above the glittering river that drifted hazily beneath the Spanish moss-covered live oaks.

Amidst the warmth and peace, I made a decision. I would throw in my hat with the U.S. Navy.

"The navy? Kathy, you're going to work with the Naval Research Lab? You, who hate the Cold War? Where?" pried my traveling companion.

"The Arctic," I answered, with some unease.

"Again?" My friend looked at me with disbelief.

The decision surprised even *me*. Perhaps Peter Vogt from the Naval Research Lab was just as surprised when I accepted his offer to collaborate in his research, using side-looking sonar in the Norwegian-Greenland Sea. For me, this was a strange yet welcome gift. As before, I was tired of writing proposals to the National Science Foundation, and this opportunity would provide near certain funding for a couple of years.

Before the end of the Cold War, a research scientist pretty much had to get in league with the U.S. military in order to receive U.S. funding for serious work in the Arctic. Normally, this would not have appealed to me, yet the military had a chain of command that was fairly straightforward compared to the labyrinth within the National Science Foundation. When I worked with the Office of Naval Research, I generally found out immediately after writing a proposal whether my team would be funded. And the requirements for the proposal were minimal.

Because of world politics at the time, the Office of Naval Research had superb relations with Norway. So did I. Peter, Eirik Sundvor from Bergen, and I agreed to join together and map the mid-ocean ridge and the eastern continental margin of the Norwegian-Greenland Sea. We would do this from a tiny Norwegian ship named the *Haakon Mosby*.

The Fram Strait, which separates the Arctic Ocean from the Norwegian-Greenland Sea, is one of the most important of the deep water passages in the world because it plays a pivotal role in the exchange of water masses on Earth. Half of the Fram Strait is covered with ice even during the summer. This is where we would go, and we knew it would be very cold.

It was a frigid day in late October 1989 when we boarded our ship: five Hawaiians, one Frenchman, a handful of Americans from the Naval Research Laboratory (NRL), several Norwegians, and the crew. The Hawaiians brought plenty of food with them—lots of white rice, soy sauce, and hot sauce from Thailand, the latter intended to spice up the Norwegian diet of meat and potatoes—but they didn't bring nearly enough clothes. It started to snow early on in the voyage, and it never let up entirely except during the weekly gale, which always seemed to blow on Saturdays, the day reserved for our candlelit dinners accompanied by red wine.

During the weekly storms, which would churn for days, most of us couldn't leave our bunks, much less dine or drink. Even in our bunks, we would be tossed up and then slammed back down onto

Lifeboat on board the Haakon Mosby, *October 1989 in the Norwegian-Greenland Sea*

the mattresses. Only one of our team, Chris Jones, had a stomach strong enough to withstand the violence of the northern ocean. I have never been so seasick in my life. Despite the roiling seas, we nevertheless had to manage our way onto the fog-shrouded deck to spell one another and stand watch. Ice crystals coated everything. The cables and railings were covered with a white, glistening frozen crust.

In these conditions, we worked on the open deck, trying to wrap our frozen fingers around little nuts and bolts. Once, when we recovered the SeaMARC from the ocean (which was considerably colder than the freezing point of freshwater), we had to replace its Plexiglas shell, which had shrunk about one inch all the way around because of the intensity of the cold. The shell just popped out of its rivets into the sea. We spent an entire night and day, through the constant darkness at 78 °N latitude in November, trying to redesign the instrument, hellishly drilling holes and screwing in new nuts and bolts.

It may seem odd, if not downright absurd, that this Arctic expedition was conducted in November. Yet there was a simple reason for the inopportune timing. Our team was the last on a long list of researchers signed on to use the SeaMARC, and since we were an international collaboration, our expedition carried relatively little importance to the U.S. ship and instrument schedulers. Basically, they bumped us to November because another scientific team wanted to work off the coast of New Jersey in September, our originally designated time slot.

Traditionally, we went into Longyearbyen, Spitsbergen, during the middle of each of these trips mostly to rest from being smashed to bits by the waves, but also it was an opportunity for the Norwegian sailors to buy alcohol (taxes were really low there) and to conduct ship repairs.

At the time of my first visit to Longyearbyen in 1980, when I came to meet the icebreaker *YMER*, I had just stumbled off of an expedition using the *Alvin* submersible to explore the Oceanographer Fracture Zone in the North Atlantic Ocean. Now that visit felt like hundreds of years ago, a world away.

Longyearbyen was founded in 1905 and is named for the American adventurer and entrepreneur, J. M. Longyear, who was the first person to locate coal in this region. Soon thereafter, Spitsbergen was designated a protectorate of Norway, which allowed any nation to set up economic ventures on the islands. Norwegians and Russians arrived to mine the coal deposits. The Russians set up a village named Barentsburg and filled it with thousands of people. There was a kind of friendly competition between the two countries. The Russians claimed that the ancient artifacts found on the island chain suggested that Russians had visited these distant cold shores long before the era of the Vikings. The Norwegians were never really comfortable with these claims, but the artifacts are nonetheless there in a museum in Barentsburg.

During the Cold War, there was really not much need for either Norway or Russia to continue to mine coal in remote Spitsbergen, but it was a convenient spot located near the submarine cat-and-mouse games. Russian submarines went to the Atlantic through the passage known as the Bear Island Trough before they turned to the south where Western submarines conducted surveillance on them. In effect, Spitsbergen acted like a northern monitoring outpost for both sides of the Cold War, not overtly military but a strategic location nonetheless. In fact, it was forbidden to have any military presence on Spitsbergen at all. Instead, it was open to scientists from all nations.

Soviet oceanographers started to operate out of Barentsburg in the early 1960s, and Soviet geologists were sent routinely to these islands because the geological formations there contain nearly the entire record of the earth's history, from 500 million years ago to the present.

In 1980, coal dust covered almost everything in Longyearbyen. The Norwegians tried valiantly to bring color to their homes and company buildings, and they were also kind enough to send members of the Spitsbergen Coal Company on winter vacations to Thailand. These inducements, together with the tax-free arrangements made with the government, drew people to live in this harsh environment. I am not sure what enticed people to the Soviet-Russian town of Barentsburg. Perhaps they didn't have much incentive outside of a small stipend. The coal miners were mostly from the Ukraine, the Soviet center for coal mining, and they were very tough people.

The Soviets went so far as to create a special currency for use only on Spitsbergen. They called it the Spitsbergen ruble, and it was, of course, useless anywhere else. One Russian friend of mine told me about parties held by Russian coal miners who were finally returning to Murmansk after many long years mining coal in the dark of the frozen winters. During these departures, the travelers would gather their rubles and toss them like currency confetti into the sea. Cascades of Spitsbergen rubles filled the air of the Isfjord and doubtless coat the seafloor on the route from Barentsburg to Murmansk.

It wasn't until the mid-1990s that Longyearbyen started to change. The Cold War ended, reducing the already diminished need for coal, and the towns started to die.

Nevertheless, Norway did not want to lose hold of its protectorate and gradually opened Spitsbergen to tourists. Today, it is difficult for a scientist to find lodging because all of the hotel rooms are booked long in advance. It is likewise difficult to book a ticket on the now popular flights from Tromsø.

I made several trips to Spitsbergen throughout the 1980s, while on five different expeditions. Sometimes when the ice was far to the north, we could sneak around the northern tip of Svalbard (as the Norwegians call Spitsbergen) and into the Arctic Ocean proper. During one of these trips, we brought the SeaMARC II sonar to 82° N, slipping through the greasy, slush-covered sea into a jigsaw landscape of pancake ice. These journeys were beautiful, when the sea was calm and glassy, tempered by the surface ice. You could hear the subtlest changes in the atmosphere, the winds, the ice cracking, or the slush giving way with a swoosh, while the sky reflected a mirror image of our vessel into the ocean underneath.

As with my first Norwegian expedition, the Norwegians typically ran their ships with a crew of nine. They were remarkably efficient and reliable, and they instilled a feeling of safety and security. This was important. There was no British stiff upper lip, no Russian risk taking, no American caution. The Norwegians are the masters of the North. Every person on our ship was outfitted in a fail-safe survival suit, and we practiced ocean rescues at least twice on each cruise.

Our crew was also a very close-knit bunch. The captain was not much older than anyone else, and the first mate and deckhands all worked evenly and fairly. After years of sailing with them, my appreciation for their tight teamwork was sealed forever on one of my last voyages. We were headed along the exquisite inside passage back to Bergen, and on the second day, south of Tromsø, we entered a very narrow corridor that threaded between tall mountains towering on either side of us. Without warning, the ship's engines slowed and stopped. I came out onto the deck to see what

had happened. Was a fishing boat in the way? Were we experiencing mechanical problems?

I looked out to starboard, and toward us putt-putted a tiny
wooden boat manned by two people, an old man and woman
dressed in red and gold rainwear. The boat was varnished spotless,
and its pointed, Viking ship-like shape edged closer. Suddenly, the
captain raised his arms, and the old couple responded with a wave
and a shout. They pulled up alongside and opened up the hatch
beneath their feet. Soon, I heard shuffling sounds from on deck,
and I turned to see the crew loading cardboard boxes that were
hoisted over and down to the old couple's boat. These were cartons
of liquor, a smuggling operation in action. Cheap liquor was purchased on Svalbard, stored in the hold of our ship until we met up
with the parents of the captain in the quiet waters of the inland
passage, where the alcohol was transferred.

The captain waved good-bye to his parents, and we resumed our
journey. I watched the old couple heading off silently into the distance toward an isolated island. Afterward, I learned the whole
story. The entire crew temporarily stored its liquor with the captain's parents, who lived some distance from Bergen. As Christmas
approached, one crew member at a time drove from Bergen to the
north to retrieve his or her stash. The authorities would never
know about the excess quantities, and the crew would not be taxed
for their indulgences during this celebration of light in the dark
winter world of Norway.

My expeditions with the NRL turned into some of the most
relaxed voyages I have ever experienced, in spite of the crushing
weather. During one expedition, when the weather almost cooper-
ated, we dressed up for Saturday dinners, which paradoxically
helped us to unwind because it was such a contrast to the work we
were doing on deck. After dinner, we would play guitars, sing, or

juggle, all while bracing against a heavy swell or crashing waves, bathed by the warmth of candlelight and wine. We were tellers of tales, shanty singers, deck workers, philosophers, mathematicians, artists, and manual laborers, sharing a brief but elegant respite from the frozen stormy seas. I would choose these comrades to sail with on any expedition, and this is why I kept returning "with the navy to the North."

Over two expeditions, we mapped about 45,000 square miles of the Arctic seafloor. We sailed up the Knipovich volcanic ridge, which we learned was a combination of an immense strike-slip fault (not unlike the San Andreas Fault) and a nearly new mid-ocean ridge, where lava erupted over harmonically spaced intervals along its axis of spreading. We traveled as far north as we could go, to and around the ice edge, inching bravely ever northward. Along the continental margin that separates the Barents Sea from the Norwegian-Greenland Sea, we found evidence of mud volcano eruptions and of the venting of methane into the ocean above. We also searched for seafloor traces left by cascading winter water, which is thought to form on the Barents Sea continental shelf when ice develops and salt is leached out. Salt concentrates below the ice and increases the density of water to the point where it sinks rapidly to the shelf floor. This is thought to create gigantic annual spills over into the Norwegian-Greenland Sea beyond, the driving force of the global ocean conveyor belt—that grand system by which warm ocean water is exchanged with cold ocean water, keeping the northern regions of our continents relatively temperate.

Despite our efforts, we were not successful in imaging these corridors of deepwater formation. Where were they? We were not sure. We would have to return once again.

BAIKAL

Five years had passed since the conference in Moscow, and nothing new had happened between the Soviet Union and the United States, except that Leonid Brezhnev was gone and Mikhail Gorbachev had inherited the great, grand, ailing country. Chernobyl had blown up in 1986, spewing its insidious radioactive debris over Europe. The Soviet citizens were rising up in arms. Soon, despite all the political pundits' predictions to the contrary, the Berlin Wall would crumble and fall. The budding and powerful grassroots environmental movement in the Soviet Union was helping to shift politics toward openness.

On the cusp of the change, the Soviet Academy of Sciences and the National Geographic Society asked me to return to the Soviet Union. In September 1989, I was bent over my desk at Lamont, putting the final touches on the plan for an upcoming expedition to the Norwegian-Greenland Sea and the Arctic Ocean. I was one of the chief scientists, and there were challenging problems to solve. We had to move tons of sonar gear from Hawaii to Spitsbergen, ship twelve people from all over the world to that remote spot, coordinate the U.S. contingent with our Norwegian shipmates, and prepare for a tough trip in total darkness at 80° N. I was caught in the nagging problems of moving unbelievably expensive equipment to unbelievably remote parts of the world when the phone rang.

Emory Kristof was on the line from the National Geographic Society, and true to form, he had a new expedition in preparation, and he wanted me to lead a portion of it. His new quest was something that I could not resist. The Soviet Academy of Sciences had extended an invitation to the National Geographic Society to organize the first major visit by U.S. scientists to the deepest lake in the world, Lake Baikal. Since Baikal is almost a mile deep, it would take technology developed by oceanographers to penetrate and chart its depths. What made it even more enticing was that Emory had already promised the Soviet Academy of Sciences that we would discover hydrothermal vents. How could I decline such an invitation?

Emory had pioneered a sharing partnership between the Soviet Union and the United States, whereby each country contributed scientific equipment for use in deep-sea research. He recognized the opportunity that the marvelous, spacious, and technically superior Soviet deep-diving submersibles presented for advances in the world of deepwater exploration. These submersibles, known as the *Mirs,* were the main tools later used to film both the first IMAX movies of the *Titanic* and the Hollywood blockbuster film by the director James Cameron. Because of Emory's inventive energy and creativity, and because I felt stifled by traditional research, I decided to join and do my bit to strengthen ties between the Soviet Union and the United States—but not before I completed our grand mapping exercise north of far distant Spitsbergen. Once that was completed, I turned my energy toward the pending trip to Russia's vast interior.

Lake Baikal is in one of the most remote regions of central Siberia, 200 miles north of Mongolia, near the town of Irkutsk. It is situated over an area marked by a continental rift that is spreading apart at the rate of about an inch a year, thereby increasing the lake's volume. Nobody is quite sure which tectonic plates are involved in the formation of Lake Baikal. It could be cracking apart in response to the effect of the Indian Plate crashing up into Asia about 60 million years ago. This is the way in which some oceans form, through continental rifting and spreading. This process may also stimulate the upwelling of volcanic basalt deep into the rift valley. Because basalt is heavier than the neighboring continental rock, the rift starts to sink and, over time, collects water. During the cracking and stretching of the crust, water from the newly forming lake seeps down through fissures and faults in the crust, where it is heated by contact with the hot rock (basalt) below. The hot water leaches large quantities of minerals from the magma before spewing back out the vents into the lake. Through time, as the rifting continues, the lake will grow into a sea (like the young Norwegian-Greenland Sea) and then into an ocean (like the Atlantic Ocean, which began rifting North America away from Africa more than 160 million years ago).

Among scientists in the geological community, there was a heated debate about whether Lake Baikal was rifting and whether it was volcanically active. The Soviet Academy of Sciences wanted us to test these theories. They hoped that we would find evidence of heat (a popular topic at the time), which would require us to find hydrothermal vents on the bottom of the deepest lake in the world. This seemed an awesome task. I had no maps; in fact, no one had ever constructed a decent echosounder chart of the lake.

Lake Baikal is the world's largest body of freshwater—395 miles long and 5,380 feet deep—and it contains about one-fifth of the world's freshwater, excluding the frozen water locked in the Antarctic ice sheet. It also contains an estimated 1,700 species of animals and plants, many of which are found nowhere else in the world. Freshwater seals and sponges are among its unusual inhabitants. Living conditions have proven so favorable that many distinct kinds of organisms thrive on unprecedented scales. Many of the animals strangely resemble marine (ocean) species. How did these creatures evolve in this freshwater lake? Was it once connected to the sea, or had life evolved separately right here? This question was of incredible biological significance. I was eager to unlock the mysteries of this gigantic body of water, to be among the first to map its impenetrable bottom.

In June 1990, the National Geographic Society, Tokyo Broadcasting, and a bevy of scientists descended upon an Intourist hotel in Lystvianka near Lake Baikal. We brought with us two tons of equipment, including ten cases of Hershey's M&Ms, a trunk of medical supplies, and another trunk of prophylactics (at the request of the director of the Limnological Institute). Our instrumentation included two ROVs (remotely operated vehicles), a towed search vehicle that contained one Benthos camera with a 100-foot film cartridge, a Sea-Bird Electronics CTD to measure conductivity-temperature-depth, and a 12-kilohertz Benthos pinger that would provide sled-off-bottom height information. We would navigate using three Magellan Systems Pro-1000 handheld downlink receivers and navigational computers. This operation was simplified and augmented by a special Magellan satellite software package that permitted upgrading the accuracy of each field unit by comparing it to a reference land-located unit. We rigged up these systems across all of the watercraft and added a satellite phone link in our hotel command center. This was the first time that the global positioning satellite (GPS) system had been used within the Soviet Union, and it evidently was so impressive that the U.S. ambassador paid us a visit.

Perhaps the science intrigued him, but we wagered that he seized the chance to make a free and private phone call to the Unites States.

I had allotted five weeks in Siberia for the expedition. The first week we organized the equipment and established the necessary diplomatic protocols that such a landmark scientific exchange program demanded. After meeting most of the Soviet scientists in Siberia, we were given permission to begin. As chief scientist, I decided who would make up the team. Earl Young from the Woods Hole Oceanographic Institution was first. He was an excellent engineer who could fix anything with little more than a piece of chewing gum and a shoestring. The second was Pete Petrone, a photo technician from the National Geographic Society who would create a high-tech color photo laboratory in the hold of our decrepit vessel, the *Vereshchagin*. Thirteen years earlier, Pete, Earl, Emory, and I had participated in those first submersible dives to the Galápagos vents. Emory also brought along the National Geographic engineers Mike Cole and Keith Moorehead. Vladimir Golubev, a geophysicist from the Institute of the Earth's Crust in Irkutsk, and Gregory (Grisha) Bobylev, a first-rate translator from the Baikal Limnological Institute, were also assigned to our team by the Siberian Academy of Sciences. Vladimir had years of experience taking heat-flow measurements in Baikal and claimed that he could locate areas similar to open ocean hydrothermal vents. Unfortunately, navigation errors and faulty equipment limited his progress. However, his perseverance provided the baseline information that allowed us to narrow our search region from a vast area to one much smaller and more manageable, some 50 square miles.

Although the team was assembled and most of our equipment was on hand, we were far from ready to get on the water. Pete had to fly the thousands of miles back to Moscow to retrieve essential equipment that Soviet customs had misplaced. We also had to secure another ship on the lake, in addition to the *Vereshchagin,* that could support and launch a *Pisces* submersible. The *Pisces* submersible is a Canadian-built unit that predated the Finnish-manufactured *Mirs.* Because the *Mirs* were being used in the Pacific Ocean at the time, we were limited to the use of the Soviet-owned *Pisces.*

Meanwhile, I took in the surrounding landscape. Even though the Siberian lands and people were beautiful, I was dismayed by the poorly built houses that I saw everywhere. Every structure was marked by shoddy workmanship and a dangerous lack of basic maintenance. Airfields were filled with airplanes and trucks that would never run again. I couldn't help wondering why the CIA thought that the Soviet Union was such a formidable foe—beyond its arsenal of nuclear weapons—when half of the country's equipment was broken and rusting on site. When I asked our translator about this sorry state, he explained that the Soviet system did not reward anyone for personal effort. In fact, he added, the Soviet system persecuted people who attempted to take pride in doing better work. There was no incentive for maintaining or repairing much of anything.

Once we collected all of the equipment, we left port and steamed the full length of Lake Baikal to the northernmost corner called Frolikha Bay. Baikal's water is the clearest in the world. The impregnable taiga forests and Mongolian steppes border the lake and former home of Genghis Khan. Hundreds of species of wildflowers, including Siberian irises, Asian lilies, and orange globeflowers, and wild apples and cherries speckled the shoreline. Below us, we could see to depths that surpassed 90 feet. We would survey in water depths of 1,200 to 2,100 feet, but because the walls of the lake were so steep, we were very close to land. Everyday we stopped to fish, to replenish the onboard food supply, and to seize the chance to wander along the shoreline and into the forest.

On one hand, this was a wonderful alternative to spending months at sea far away from any sight of land. On the other, the ship was filthy. I had a cabin down below where a half an inch of dust covered everything. Flies propagated by the hundreds and buzzed like a seething black carpet over my bunk. I was disgusted, wondering how I could ever lay my body down on such a place. I solved the problem by borrowing the sleeping bag of my colleague, Ralph White, who was stationed back at the hotel. For some reason that I shall never know, Ralph's sleeping bag disappeared, and to this day he will not let me forget it.

Flies were the least of our problems. Most of us were battling stomach upset doubtlessly caused by the putrid water in the hotel. There was no toilet on the ship, just a dirty hole in the floor. Our cook, an alcoholic, spent most of her time scrounging anything containing alcohol (including perfume) instead of preparing meals. Fortunately, we had brought our own provisions and a microwave. Pete not only set up a photo lab but created a functional onboard kitchen.

Then the adventure began. Because no map existed of the lake floor, we were left with very old and relatively useless depth charts. No oceanographer would lower equipment into such depths to make a new map. We opted to use *Vereshchagin's* echosounder, except that it was stationed on the bridge and scientists were never allowed to use it. We devised a means to post one of our team to the bridge, to expedite communications between the English-speaking scientists and the Russian captain. It worked. We also rigged together a proper echosounder data recorder in the lab. I arranged a simple grid search pattern, and from the depths measured along the tracks, I constructed enough of a bathymetric chart to allow us to navigate the camera sleds. Even so, the ship could not make refined course changes and could not travel at slow speeds. We were forced to tie up a smaller ship to the *Vereshchagin* and deploy it separately to take data readings and measurements that would guide our navigation. Either Pete or I would determine a course change, Grisha, our translator, would run up to the bridge, the captain would shout out the window to the captain of the smaller vessel, and he would make the course change. It was a long chain of command and communication.

After a week of photographing mud and measuring subtle temperature anomalies, we chanced upon a temperature spike of 0.1°C, a very significant change in the deepwater environment. We were close to hydrothermal vents. At the same time, the camera showed mats of luminescent white bacteria covering an area the size of two football fields. We marked the location using the GPS system and carried out three additional depth transects to further delineate the boundaries of the hydrothermal vent field. In addition to the thick mats of bacteria, which signified that hot water laden with sulfur was bubbling up from below, we detected numerous white encrusted sponges. Coiled gastropods and whitish translucent amphipods clustered among the sponges and on the sediment at the edge of the bacterial mat. We were jubilant.

Following the discovery, we sent Earl Young by hydrofoil back to the Limnological Institute in Lystvianka with copies of the photographs. There, 400 miles away, Emory Kristof and Mikhail Grachev, the director of the institute, anxiously awaited news from our team. It had taken weeks to ready one of the *Pisces* subs for work in freshwater; freshwater is lighter than saltwater, and a submersible would sink like a stone unless the buoyancy were changed.

Once the sub was deemed usable, it was towed by barge up to Frolikha Bay. While we waited for the subs to be mobilized, Emory

Emory Kristof sacked out in the laboratory sleeping quarters of the Papanin. Keith Morehead watches.

and his ROV team invited me to join them on their tiny ship, the *Papanin*. This little vessel, hardly a ship at all, was assigned to map the obscure crevices in the rift valley walls, which were inhabited by deepwater sponges. They would film the hidden world of the only existing freshwater seals as well. I was one of only three women on the ship, and because there was so little space, we bunked in the mess hall. My bunk was on top of the dinner table. The men camped out in their tiny work quarters in the aft part of the ship. All the electronics and ROV pieces were thrown together into crates that they cobbled together into makeshift beds. Emory, Mike, and Keith would sack out after drinking several beers and shots of vodka.

The *Papanin* may have been cramped, but it did have a spectacular cook. The Russians claimed that she was a witch, which she admitted as well. She said that she specialized in the black arts, including death and dismemberment. The Siberians tend to be avid believers in the occult. In addition to the cook, a marine biologist on board was convinced that she could cure fatal diseases with the use of her magnetically powered hands. Mike, the engineer, gave the cook a wide berth; when he boarded the ship with Olga, his Russian girlfriend and our translator, she greeted them with a "knife across the neck" hand movement. Olga and Natasha, another Russian friend, added their own superstitions about the cook. Olga implored me, "Please, don't leave the ship. We are sure that your presence is protecting us from the cook's evil powers."

Nevertheless, I had to return to the *Vereshchagin*. I had with me a new heat-flow instrument developed by Woods Hole for use on a submersible. It would allow us to measure precisely the amount of heat venting from the bottom of Frolikha Bay. The National Geographic team attached a video and still cameras to the submersible's other manipulator arm. The pilot received course directions from the Magellan Pro-1000 operator at the surface. The surface operator also used a Ferranti Trackpoint II underwater navigation system to fine-tune the piloting. Detailed measurements revealed that the vent area was laden with minerals and that the water temperature was 24°F warmer than the surrounding lake water. The area registered more than 900 times the normal heat flow of the earth.

Some 85 percent of the photographs suggested to us that similar concentrations of organisms or sponges would *not* exist outside of the hydrothermal vent field. Lake Baikal's native ecology is unique due to the number of local species that are similar to saltwater counterparts. How did these life-forms evolve when the lake was located thousands of miles from the sea? Perhaps the lake was once connected to the ocean or perhaps chance led to the development of life-forms astonishingly similar to those in the ocean. Just

National Geographic Team, Lake Baikal, Russia, 1990.
PHOTO: *Ralph White*

Pisces diving operations, Lake Baikal, 1990
PHOTO: *Ralph White*

Lake Baikal's underwater landscape and life-forms, ROV dive, 1990
PHOTO: *Emory Kristof*

how these life-forms fit into the biogeographical arena of hydrothermal vent communities remains a giant unsolved puzzle.

The discovery of deep-freshwater vents in Lake Baikal suggests that the Baikal Rift may be very volcanically active and that the rift is gradually widening into a new ocean. In addition, the vents provide an opportunity to investigate the evolution of creatures not dependent upon the sun for their existence. The biological implications of this are significant.

I feel fortunate to have participated in this expedition, to have seen firsthand the magnificent wonder of a newly discovered underwater environment, and to have played a role in unraveling its secrets. Lake Baikal's mysterious treasures reveal the earth's incredible diversity, created over millions of years of evolution. However, what took nature millions of years to create can be

destroyed in ten by humankind. Siberian scientists and environmentalists hope to educate the Western world about Lake Baikal so
that an international ecological center can be established on its
shores, protecting the life within from the onslaught of pollution
and destruction.

Our research focused the world's attention on this unique environment. Lake Baikal is now considered to be "one of the seven
underwater wonders of the world." I am proud that I contributed
to this important discovery, and I remain hopeful that Lake Baikal
will be protected. I left Siberia with a touch of sadness because I
was leaving behind its people and its beauty. We had made extraor-
dinary scientific discoveries, but the expedition was also an important step in breaking down the Cold War barriers that had existed
between our countries for decades. I am privileged to have participated in that as well.

25

RAPID RUSSIAN RESPONSE

Strangers in the night
—Kaempfert, Singleton, and Snyder

Somewhere in the Norwegian-Greenland Sea, I awoke in the middle of the night from a strange, image-filled underwater travelogue of a dream to a very dimly lit cabin. My roommate, Lena, slept in the bunk above me. Beside me rested our mounds of winter clothing mixed with teacups, a vodka bottle, some chunks of chocolate, and an orange that I had spirited away from the mess hall that afternoon. The subtle sounds of a saxophone worked their way through the porthole and the narrow cracks around the cabin door. The melody of *Stranger in the Night* swirled in and around, and I thought of the lonely Russian captain playing to the seagulls as we took the inky swells of the ocean one sway at a time. We were aboard the research vessel *Professor Logachev,* far away from its home port of St. Petersburg.

It was 1996. And although we hardly knew it, it was summer. We were back in the cold Norwegian-Greenland Sea carrying out the world's first trilateral U.S./Russian/Norwegian expedition, searching for gas hydrates and mapping the volcanically active Knipovich Ridge. Since 1990, I had helped to organize five ocean-going expeditions with our Soviet-Russian colleagues. This success was in sharp contrast to previous years, when there was little, if any, permissible contact with our Soviet counterparts. With the fall of the Berlin Wall and the end of the Cold War, the world was completely reordered, and people, countries, and cultures that were previously estranged from one another now worked in close collaboration.

In early 1991, just after our successful exploration of Lake Baikal, I moved to the Hawaii Institute of Geophysics during the first half of my sabbatical year from Hunter College. There I worked with the SeaMARC II group to process side-looking sonar imagery that we had obtained some years earlier from the Norwegian-Greenland Sea. During one balmy Hawaiian winter day, I received a phone call from Peter Vogt, Lynn Johnson, and Chris Jones, who were planning a "rapid response mission" with the Naval Research Lab (NRL) to a suspected volcanic eruption site on the Reykjanes Ridge in the Atlantic Ocean, just south of Iceland. Vogt knew of my previous success with the Russian oceanographic community in Lake Baikal, and he wanted my assistance in negotiating the use of the Soviet *MIR* submersibles for this proposed expedition.

The prospect of searching for an active volcanic eruption on the seafloor in the cold reaches of the North Atlantic was very enticing. However, I was scheduled to move to Washington, D.C., in late 1991 to develop the Environmental Defense Fund's new initiative, Arctic at Risk. At the time, I was eager to make the transition in my career from geophysics to environmental security, but I agreed to send a fax to Anatoly Sagalovich, the chief of the *MIR* diving team in Moscow, to inquire about the availability of their subs for the mission. Two days and many time zones later, we received his answer: yes. In July 1991, the Soviet research vessel

Keldysh would be available along with the *MIR* submersibles for investigation of two suspected lava flows on the Reykjanes Ridge. Because the *MIRs* were the only research submersibles available in the North Atlantic during that summer, and the NRL research team had been invited to dive on this site at nominal cost, we submitted a proposal to the National Science Foundation to support this unique and affordable opportunity to investigate, verify, and sample firsthand the region on the Reykjanes Ridge. Although the expedition was called a rapid response mission, it had taken two years to contract a submarine. That's how things are in oceanography; it takes many months to plan an expedition.

The *Keldysh*'s schedule was very complicated. It was already committed to a series of dives in May and June 1991 at the hydrothermal vents discovered by an earlier *Logachev* expedition at the Mid-Atlantic Ridge. In addition, the *MIRs* had been booked by the Soviet navy to dive on the sunken Soviet submarine *Komsomolets* to assess radionuclide contamination that may have seeped from the craft into the surrounding bottom of the Norwegian-Greenland Sea. Afterward, they were committed to participate in a project at the *Titanic* site with the IMAX Film Corporation and the National Geographic Society. For a nominal sum of $40,000, we were told that we could squeeze one and a half weeks out of their busy summer schedule to carry out our investigation of the Reykjanes Ridge. The *Keldysh* was scheduled to arrive at the port of St. John in New Brunswick, Canada, on July 8 and would be ready for departure to Rotterdam, Netherlands, on July 12. The National Geographic Society had funded the ship time necessary to get the *Keldysh* back to her home port of Kaliningrad by way of Rotterdam. Our one week of dives on the Reykjanes Ridge, midway between the two ports, would fit into the ship schedule nicely.

Somewhere during the planning stages, the scheduled ports changed to Reykjavik, Iceland, and Copenhagen, Denmark. I don't remember why this happened, but it was not unusual during Russian expeditions; their planned routes were often subject to financial constraints when they worked with the West.

The U.S. team consisted of Chris Jones and Lynn Johnson from the NRL, Bruce Appelgate, my colleague from the University of Hawaii, and Roger Buck and myself, representing Lamont-Doherty and Hunter College. After having endured a very long voyage from Hawaii to Reykjavik, Bruce and I had anticipated a rather quick turnaround from landing at the airport to boarding the ship. Instead, when we arrived in Reykjavik, the *Keldysh* was not yet in

The U.S. team: Roger Buck, Lynn Johnson, Bruce Appelgate, Kathy Crane, and Chris Jones

port—and wouldn't be for several days. Given this unanticipated period of free time, the U.S. team decided to cram into a rented Lada (a small but reliable Russian car) and traverse the whole of Iceland. If we were destined to fail in our mission to map the Reykjanes Ridge in the ocean, we would at least have a chance to visit the Reykjanes Ridge where it ran aground in Iceland. The possibility of failure was always high when working with the Soviets, because there were so many unforeseen factors that could influence their arrival in port. This time was no different.

Squeezed together like sardines in a can, we bounced over volcanic terrain, used dead reckoning to locate the volcano Krafla, stayed at farmers' houses in tiny towns, and took the opportunity to plunge into the hot geothermal pools that dotted the landscape. By the time we returned to Reykjavik, the *Keldysh* had arrived, but its crew was in revolt. Not only had the NSF funds *not* reached the ship (apparently, the director of the Shirshov Institute had taken the money for something that he had deemed more important), but the crew hadn't been paid in months.

During the planning of the expedition, I had worked to secure NSF funds for the Shirshov Institute. Anatoly Sagalovich encouraged me to send the funds directly to his special *MIR* account in Germany and not to the institute's director. However, the accounting department at Columbia University, which coordinated cash transfers among the NSF, Lamont, and the Russian institute, insisted that the funds be sent directly to the director, with the instruction, "Directors must deal with directors."

With foreboding, we sent the funds to the Shirshov bank account, and subsequently, the funds were "misplaced." This expedition was the first Soviet expedition funded by the NSF, and now the money had disappeared. Sagalovich was nowhere to be found, at least not in Russia or in Iceland. Instead, he was in the United States trying to raise additional funds for future collaborative work with the IMAX film company at the *Titanic* site.

The *Keldysh* crew was paralyzed. The U.S. team climbed aboard and was led into the chief scientist's cabin, which was luxurious by any standards, and asked to sit down for a ritual of vodka drinking. Chris Jones, our official drinker (it was 9 A.M.) was critical to the success of the ritual. Shortly into it, we learned that the situation was even worse than we had imagined. The *Keldysh* could not pay the ship's agent in Iceland. Instead, the Soviet team seemed content to drink vodka with a sense of gray fatalism. This went against every fiber of my nature. I asked, "Why don't you call Anatoly?" There was no answer.

I decided to take action. I would have to call in favors to extract us from this mess. The first person I called was Ralph White, a friend from Los Angeles. We had worked together in Baikal, and he was a close colleague of Emory Kristof. At various times in his career, Ralph had been an underwater cinematographer, as well as head of the Los Angeles Police Department's rifle squad. Now he was the Western liaison for the *MIR* group. Sure enough, Ralph was able to locate Anatoly in the United States and also connected me with Joe MacInnis, the Canadian doctor who had participated on the Lake Baikal expedition. MacInnis was involved in the venture to film the *Titanic* with the *MIRs* together with IMAX in Montreal, Canada, and he had set up a separate bank account in Canada for *MIR* activities. Joe, by way of IMAX, asked me how much money I needed to get us out of port, to pay the crew, and to conduct the expedition. I told him, "$40,000," which was exactly the amount that NSF had granted to Lamont-Doherty for the expedition.

The very next day, $40,000 was transferred to the Reykjavik ship agent, and suddenly our operation with the Soviets was revived. I never told NSF that their money went missing somewhere in Russia, because I was sure that they would never again collaborate with such a partner. I am forever in the debt of White, MacInnes, and IMAX.

We now had an expedition to conduct. We packed up our meager belongings, returned the Lada to the rental car agent, boarded the *Keldysh*, left port, and dove to the seafloor. However, despite having overcome these frustrating obstacles, our search proved nearly fruitless. We were unable to locate any signs of recent volcanic activity outside of a few glimpses of fresh lava. Only well after the expedition, when our team had returned to the University of Hawaii, did Bruce Appelgate, Lynn Johnson, and I discover the problem. Errors in the original earthquake locations had been transferred from map to map, resulting in a collection of dives that missed the targets.

Although we did not locate the volcanoes, the diving operations were conducted with utmost professionalism. The expedition also had its entertaining moments. I remember how strange it was to be diving in a Soviet submarine on an underwater ridge that was the common hunting ground of adversarial U.S. and Soviet nuclear subs. I imagined a U.S. submarine coming alongside and questioning us because we were Soviet, although there were two Americans inside the *MIR 2*. Once, I dreamed that we yelled through our radio to a neighboring U.S. nuclear sub, "Hey, leave us alone. There are two Americans in here. Don't you know the Cold War is over?"

Another episode sprang from cultural nearsightedness. Miraculously, the *Keldysh* had a doctor in its service, and he routinely examined all divers before and after the dives. The exams, however, consisted of little more than old-fashioned taps on the knees, a stethoscope to the chest, and induced coughing episodes. We were lined up one by one outside of a closed curtain, but we could hear everything between doctor and patient. For the men, the routine was the same, but when I stepped behind the curtain, the doctor used the little English he knew and blurted, "Take off your clothes."

Roger Buck loved this line, which he said came directly out of the movie based on Milan Kundera's book *The Unbearable Lightness of Being*. When we passed the doctor on our way to climb into the submersibles, Roger shouted out gaily, "Take off your clothes!"

When we returned from the abyss, we followed the Russian habit that would help us recover: The men went to their *banya*, the steam room, and Lynn and I went to another. Bruce, Chris, and Roger claimed that they loved having their backs beaten by birch

Inside the Russian MIR 2 submersible, with Chris Jones, and Genia, the pilot, on the Reykjanes Ridge, south of Iceland, at a depth of 700 meters

branches, something the men did to one another in their *banya*. Lynn and I had a more sedate experience. I still wonder about those birch branches.

The *Keldysh* expedition had its challenges. However, we quickly learned to operate both within and outside the system, to accommodate the peculiarities of oceanic exploration with the Russians. The *Keldysh* expedition cleared the way for many subsequent expeditions, which would be filled with laughter, song, good food, and camaraderie. The logistical success of the Soviet rapid response mission in 1991 opened the door to another expedition with the Shirshov Institute's *Mendeleev* in 1993, but one with a far different outcome. It would soon become clear that our basic research missions with the Russians were always more successful when our governments did not interfere.

26

SOOT

On November 9, 1989, the Berlin Wall fell, people stormed out of Eastern Europe, and the Soviet Union began to crumble. What remained was an extraordinarily decrepit and polluted world that festered in the midst of heartbreaking beauty. Statistics revealed that the cancer rate in the East was twice that in the United States. People were suffering severe health effects as a result of the enormous pollution generated by the Soviet military-industrial complex, and the contamination from nuclear and hazardous waste extended deep into the Arctic. For decades, corruption in the Soviet bloc had routinely rewarded mediocrity and cover-up. The debilitation and loss of life that Soviet citizens had experienced was staggering. The Soviets themselves did not comprehend the magnitude of the devastation; during the Cold War, the need for self-preservation overrode any consideration of environmental health and safety. I saw some of the devastation firsthand, and it so moved me that I felt compelled to help solve some of the problems. Many humanitarian organizations interceded quickly to bring Russia and the United States together, but comparatively few scientists shared in this effort, with the exception of physicists who worked on the problem of "loose nukes."

Why were so few American scientists interested in building contacts with Former Soviet Union (FSU) scientists? Why did the U.S.

National Science Foundation play such a small role in supporting the survival of FSU science? From my own experience in academic research, I had learned that many U.S. scientists were rather provincial and somewhat egocentric, believing that no other country could compete with the United States and that working outside the U.S. system would provide no benefit for individual advancement. Ironically, it was the Department of Defense that spearheaded much of the effort to fund the ailing scientific establishment of the Former Soviet Union. I would soon be brought again into their sphere of influence.

Because of my budding interest in the field of environmental security, Stephanie Pfirman and Scott Hajost of the Environmental Defense Fund (EDF) invited me to join EDF in the late summer of 1991 as a visiting scientist to develop a new program called "Arctic at Risk."

Stephanie Pfirman and I had worked for more than ten years on ships near Svalbard (Spitsbergen), sometimes operating close to Soviet territory. We knew too well that the Russian environment was extremely at risk. It was painfully clear to me that Russian industry was crumbling, machinery did not work, there was widespread pollution of the air, water, and soil, and the health care system was collapsing. Since Russia borders much of the Arctic, we decided to focus on the environmental health there; we felt this was particularly important since no one funded outside of the Soviet military complex had monitored the region for more than fifty years. No maps existed that detailed either the sources of contamination or the range throughout which these contaminants might be transported across the Arctic. Furthermore, given the dreadful relations between the East and the West, the existing maps of the Arctic Ocean floor were poor at best. The Soviets had mapped their sector, and the West had mapped its sector, but no one had managed to join the two pieces because of the Cold War hostilities.

I saw Russia as a vast region of the earth that was practically unknown to most Americans. Over the years I had developed close relationships with my Russian colleagues, but the Cold War had limited our professional interaction. Now, a political thaw was developing, and I wanted to actively engage Russian scientists in merging previous research into one data bank. However, our program at the Environmental Defense Fund had minimal funding and poor computer resources. Nevertheless, we managed to construct the first maps of contamination in the entire Arctic region, although we had little idea how we might use this information.

In 1989, Finland initiated an action that launched an Arctic nation cooperative agreement to protect the Arctic environment, called the Arctic Environmental Protection Strategy. Goals and guidelines were adopted at a ministerial meeting in Finland in June 1991. Part of this initiative was the foundation of an Arctic Monitoring and Assessment Program (AMAP), which established systematic measurement of pollutants and assessment of their effects on the Arctic environment as its primary objective.

Our team of Arctic experts at EDF wrote a grant proposal to the U.S. Department of State to develop the first atlas of the Arctic environment. The funding we received paid my salary and allowed the purchase of a hefty number of colored pencils. At the time, the Department of State needed a concrete product to present to the other members of the eight Arctic nations cooperative, and more important, its employees needed accurate information on what actually had been and was now occurring in the Arctic. The level of ignorance was very high. The polar division of the

State Department focused mostly on the Antarctic, primarily on the development of international treaties there. Our funding at EDF was tenuous not only because our program, "The Arctic at Risk," comprised mostly visiting lawyers and scientists but also because the development of a vigorous Arctic program meant that some other EDF programs would probably be cut, and that meant the likely loss of jobs.

Then we got lucky. On December 18, 1991, Joshua Handler wrote an article for the *Christian Science Monitor:* "Soviet Subs: A Neglected Nuclear Time Bomb." The article sent shock waves through the U.S. Congress because Handler revealed that nuclear reactors from Soviet nuclear submarines had been dumped in the shallow waters of the Kara Sea. This action violated the London Dumping Convention, which the Soviet Union had ratified in 1972. What would happen to U.S. fishing grounds in the Bering Sea if the reactors leaked? Would a plume of highly radioactive water be swept along the Siberian Current toward Alaska? Would Alaskans be harmed?

By February 1992, articles had been published in the *New York Times,* "Secret Nuclear Dumping by the Soviets Is Raising Fears for Arctic Waters," and in the *Boston Globe,* "Soviets Reported to Have Dumped Nuclear Waste in Arctic Waters." British television also reported that thousands of tons of nuclear waste had been dumped secretly in Russian Arctic waters for more than twenty years creating a "ticking time bomb" that threatened the whole of Europe. Independent Television News quoted Andrei Zolotkov, an engineer with ATOMFLOT (the Murmansk based organization that operated the Russian nuclear-powered icebreaker fleet), as saying that the sea off the northern coast of Russia had been used as a major dumping ground for radioactive waste for years.

By July 1992, governments were scrambling to find any information at all about the levels of hazardous contamination in the Arctic. A Russian research vessel, *Viktor Buynitskiy,* packed with sonars and scientists, was slated to depart from the Norwegian port of Kirkenes on a joint Russian-Norwegian expedition. *Science* magazine published an editorial in July 1992 on the Arctic data collection and the efforts of the United States to come "up to speed." The article reported that since May, officials at the Environmental Protection Agency and the National Science Foundation had been:

> scurrying around, searching for information to confirm or disprove the reports about Russian radiation. While the NSF and EPA view the potential environmental threat as serious, they have adopted a "prudent and cautious approach" to funding new research. . . . Some research managers say they are hoping the U.S. Navy will fill the gap by sending a ship to the Arctic. But Leonard Johnson of the Office of Naval Research (ONR) says it's probably too late in the year to organize a trip. ONR is "trying to get our act together" and prepare for a careful survey, he says. "We don't want to just dash up there and take a lot of water samples that might have no practical use." Still, the first stages of a U.S. research effort are emerging. . . . The State Department has proposed making Arctic pollution a major topic of study at the new, U.S.-funded International Science and Technology Center based in Moscow. . . . Senator Murkowski himself has scheduled a hearing on this topic.

Hearings were indeed scheduled, both in Alaska and in Washington, with invited representatives from government agencies, universities, and nongovernmental organizations. The most penetrating questions were aimed at the heads of the government agencies: "Why have we found out about this great environmental threat to America from nongovernment organizations and not from our own governmental agencies?"

This very pointed question would have greatly embarrassed any government agency director. But the simple fact remained that funding is capricious and that trying to move the government in any direction by working within its bureaucracy is nearly impossible. Information must be pried out of the government by academics or nongovernmental organizations.

The effects of the congressional hearings were profound. An appropriation of $10 million was assigned to the Nunn-Lugar Act to improve the security of Russian nuclear facilities and to ensure that Russian nuclear scientists monitored nuclear production and disposal within their country. The funds were sent initially to the Office of Naval Research to be distributed to various government agencies and universities, who together were to investigate the levels of present-day radionuclide contamination in the Arctic, especially in the Kara Sea, the site of the reported dumping. No funding was allocated to the National Science Foundation, since its polar programs were not designed to handle complex Arctic environmental monitoring issues. Moreover, it was decided that the practical aspects of monitoring contamination did not fall into the category of "scientific research."

Shortly thereafter, I received a phone call from Peter Vogt of the Naval Research Lab. "Why don't you come to work for the *real* Defense Fund?" he proposed.

It appeared that the Arctic research group at NRL would receive a couple of million dollars to search out data and develop an envi-

ronmental Arctic geographic information system (GIS) under the
new Arctic Nuclear Waste Assessment Program (ANWAP) initiative. Peter wanted me to manage the GIS elements of the program,
as well as participate on the first side-looking sonar expedition to
the Kara Sea aboard the Shirshov Institute of Oceanology's ship,
the *Mendeleev.* This was the perfect opportunity, and I answered
simply, "Yes." I would move to NRL at the completion of my one
and a half years at the Environmental Defense Fund.

Our expedition likely would be controversial. The mutual U.S.-
Russian goals were to map the location of the actual radionuclide

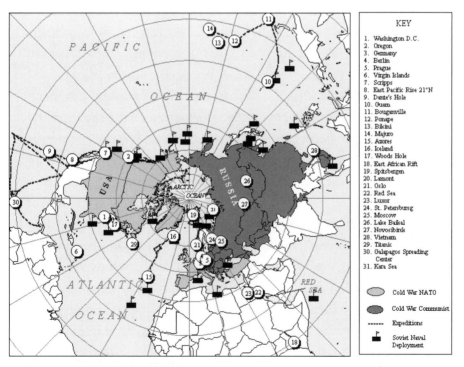

KEY

1. Washington D.C.
2. Oregon
3. Germany
4. Berlin
5. Prague
6. Virgin Islands
7. Scripps
8. East Pacific Rise 21°N
9. Dante's Hole
10. Guam
11. Bougainville
12. Ponape
13. Bikini
14. Majuro
15. Azores
16. Iceland
17. Woods Hole
18. East African Rift
19. Spitsbergen
20. Lamont
21. Oslo
22. Red Sea
23. Luxor
24. St. Petersburg
25. Moscow
26. Lake Baikal
27. Novosibirsk
28. Vietnam
29. Titanic
30. Galapagos Spreading Center
31. Kara Sea

Cold War NATO

Cold War Communist

------- Expeditions

Soviet Naval Deployment

Location Map, Places, Politics, and Expeditions

dump sites in the Kara Sea, sample the regions adjacent to the dump sites, and develop a risk assessment from the gathered data. Money was transferred from the U.S. government to the Shirshov Institute—no easy feat, based on our earlier experiences. In addition, because the program was very political, very few of the participating scientists were involved in the planning stages. The Office of Naval Research sent two individuals sanctioned by the U.S. Department of State to Moscow: Dr. Leonard Johnson, head of the Arctic Research Program, and Dr. Edward Pope, formerly of the ONR intelligence community. Peter Vogt and Chris Jones, the two expedition scientists from the Naval Research Laboratory, accompanied them to ensure that the shipboard requirements were achievable. Six months earlier, I had learned from contacts at the ABC media network that a pending conference in Archangelsk, Russia, would focus on Arctic nuclear dumping. I alerted my NRL colleagues and advised them to attend this meeting so that they could obtain early information on the situation and be poised to take part in the Arctic Nuclear Waste Program. It worked. NRL's David Nagel and Peter Vogt attended, and I joined them as a representative of the EDF. With important scientific contacts established, we hoped to take a lead role in the contaminant mapping of the Arctic—after I moved to NRL.

Our first, and only, environmental expedition was an exercise in futility. I like to call it our "Hotel Expedition," because we were never allowed to go to sea. Our frustration began once we realized that we would be excluded from the practicalities of planning the expedition. For the American side, the ONR and State Department made all the decisions, without any input or advice from either the Russian or the American scientists. However, the U.S. chief scientist Peter Vogt nearly saved us on that first mission; not only was he a well-known geophysicist, but he also spoke Russian. Peter and Alexander Lizitsin from Russia, tried as hard as they could to get the expedition going, but they were overwhelmed by intergovernmental politics.

The U.S. team included members who were experts at side-looking sonar analysis and radionuclide assessment. We transported tons of equipment to Kiel, Germany, where we were supposed to load the *Mendeleev* after it arrived from its home port of Kaliningrad. All the equipment arrived, the international teams of scientists arrived, but the *Mendeleev* did not. Every day we received the same cable message: "The *Mendeleev* is underway."

We were in port for about two and a half weeks, waiting and moving from hotel to hotel (because they had previously been booked by other people), spending the expedition funds on thirty different rooms in Kiel, Germany. Finally, fed up with the interminable delay, Chris Jones, Christian de Moustier (also of the NRL), and I drove to Berlin. The Berlin Wall had fallen only recently, and we rented rooms in a hotel in the eastern part of the city, close to the city center, near the museums with their bullet-cratered facades. The energy in the revitalizing city was inspiring. Our return to Kiel was not.

When the *Mendeleev* finally arrived, we discovered that the Russian equipment had been stolen from the trucks that had transported the gear from Moscow to Kaliningrad. This was but one of the reasons cited for the delay. Tens of thousands of dollars had been sent to Moscow, and verbal guarantees had been given to the ONR that the expedition would take place as originally planned. This was obviously not the case. In addition to the lost gear, a Russian admiral apparently vetoed the plan that allowed foreigners access to all the waters of the Kara Sea. Instead, we would be confined to only "one latitude and longitude." This meant only one point for our research effort.

It is impossible to conduct a side-looking sonar mapping expedition of the seafloor at only one point. In response, the State Department ordered all U.S. government personnel off the ship. Off came the tons of equipment, and off came the U.S. team. The total embarrassment provided a textbook example of how *not* to set up an international expedition. Those of us who were supposed to do the work on the ship fumed at the lack of commitment from both the Americans and the Russians. To this day, there has been no confirmation of radionuclide dumpsites in the Kara Sea.

During the debacle surrounding the *Mendeleev* expedition, I managed to fly from Kiel to Kirkenes, Norway, to attend an international conference on Arctic radionuclide waste. On the way, I stopped in Oslo, where I ran into Charles Newstead, an acquaintance of mine from the U.S. Department of State. Normally a jovial person, he was rather shocked to see me in Norway; I was supposed to be on the *Mendeleev,* carrying out a U.S.-funded mission. "What are you doing here? Get back to your ship!" he halfway joked, because he had no knowledge that the U.S. team had been ordered off the ship. Needless to say, he was somewhat concerned about the circumstances of our defeat.

The first month with the Naval Research Laboratory could have derailed me, but I resolved to assemble the geographic information system of the recently collected Arctic contamination data. I realized that to succeed, we needed to retrieve data already gathered by Russian scientists in their sections of the Arctic. To do this, I would have to rely upon my Russian colleagues and other international contacts, namely the scientists with whom I had previously worked. I would ask them to advise me about other trustworthy and competent Russian scientists and about other sets of unclassified data that had been published but were as yet unavailable in the United States. In exchange, we would secure funding to support Russian scientists in processing the data. Much of this work was coordinated with similar activities promoted by European countries, and this led, in many cases, to a compounding of positive results. In particular, my close ties with Heidi Kassens, my German counterpart, led to the development of successful collaborative environmental and basic research programs involving German, Russian, and U.S. scientists.

The following years led to three more Russian expeditions (to search for gas hydrates and to map mid-ocean ridges) with Russian scientists from VNIIOkeangeologia, the Shirshov Institute, the Institute for Microbiology, the Vernadsky Institute, and Moscow State University. In contrast to the *Mendeleev* disaster, these three expeditions were great successes—for scientific research, for collaboration among scientists, and for relations between Russia and the United States. After the expeditions, we brought several Russian scientists to the Naval Research Laboratory to process and digitize their data. Almost all of them stayed in my home; the NRL had no money for hotels, it was difficult for Russians to drive or rent cars, and they would have been insulted to stay somewhere *other* than my home. During that time, there were so many vodka bottles in my trash that I wondered if the CIA monitored our activities (we were only two miles from the CIA headquarters in Langley, VA). Every day, the Russians and I would fill my little Honda Civic and head to the Naval Research Lab.

The GIS project turned out to be an enormous task, and I was understaffed at the Naval Research Lab. Fortunately, I discovered Jennifer Lee Galasso (whom I knew as Jenny Lee) at a 1995 Arctic conference in Boulder, Colorado. Jenny was earning her master's degree at George Washington University, and to my delighted amazement, she was actually interested in working at the NRL on

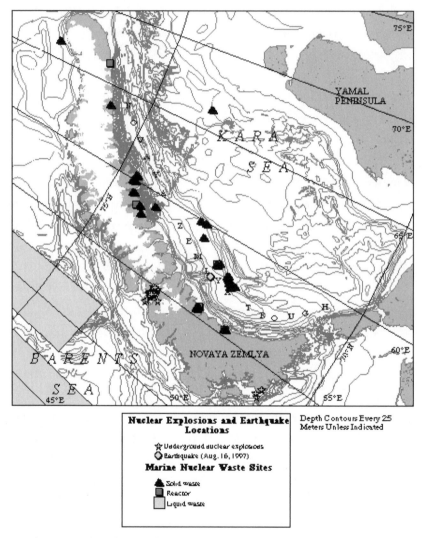

Nuclear Waste and Test Sites, Kara Sea

GIS development. I hired her on the spot. Without her help, we never would have completed our project. Jenny Lee and I assembled all the data we could gather about the physical, chemical, and biological states of the Arctic. We included new bathymetric data, sediment types, and percentages of clays, biota distributions, contamination levels of heavy metals, organochlorines, as well as radionuclides in the biota, sediments, and water, and finally, sources of these contaminants.

The developing work elicited a range of responses. Some members of the U.S. military objected to the use of the color red, because they felt it might frighten people. The British did not like the big red streak on the map that illustrated radionuclide contamination emanating from their Sellafield fuel-reprocessing center. The Russians took issue with the maps that showed contamination flowing out of the Ob River in the 1960s and 1970s. The Norwegians bristled over the highlighting of any pollutants in the Barents Sea, their prime fishing grounds. Some notable Alaskans objected to any illustration of the *Exxon Valdez* oil spill. In fact, one senator bluntly told us, "It's not there anymore. Take it off your map."

To please everyone, we would have needed to produce a perfectly blank atlas. Instead, we persevered and convinced the ONR that the color red was traditionally used by mapmakers to indicate a large quantity—not an element considered to be either "bad" or "good." The color blue indicated a smaller quantity, again, without any value of "good" or "bad" otherwise assigned to it. The result was an atlas that included an appendix, which clarified the contaminant levels that the different governments and organizations considered dangerous for their respective environments. We were surprised that, in many cases, there were huge differences in interpreting critical danger levels.

Our *Arctic Environmental Atlas* was finally published in 1999. We presented the atlas, under the auspices of the U.S. Department of State, to both the international Arctic Monitoring and Assessment Programme and to the Intergovernmental Arctic Council. After seven years of developing the atlas, we concluded that the Arctic is subject to intense environmental stresses, which have resulted and

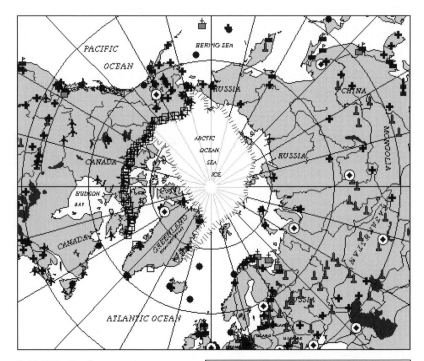

Cold War Defenses
Surrounding the Arctic

REPORTED MILITARY BASES AND DEFENSE SYSTEMS, PAST AND PRESENT

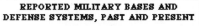

⬍ Missile Base
⬍ Navy Base
⬍ Submarine Support
✚ Other Military Support
☐ DEW Line Stations
✦ Main Air Defense
● Loran C Stations
◉ ICBM Early Warning

continue to result from the activities of the eight nations within its borders, in addition to those of many other nations throughout the world. Many of the stresses had been kept secret by the Cold War military control over the region.

Long-range transport of pollutants, resource depletion, climate change, and ozone loss are all significant environmental stresses created by humankind. Pollutants transported to the Arctic quickly make their way to the top of the Arctic food chain, resulting in unusually high amounts of toxic substances in seals, polar bears, and humans. Traditional lifestyles prevail in the Arctic and make the native peoples vulnerable to the adverse effects of toxic substances. Exposure to heavy metals is greatest where traditional hunting is practiced, where high levels of consumption of lipid-rich local food, including marine mammals, predominates. Toxic substances remain for long periods in the environment because the low rate of biogenesis in northern soils inhibits chemical turnover; this results in prolonged threat to all inhabitants.

Mercury, cadmium, lead, and arsenic have increased dramatically in the marine environment and especially in the sediments of certain regions such as the Anna Trough in the Kara Sea, throughout the Barents Sea, and in the neighboring small metal mining districts off of Greenland. Although cadmium has been used industrially only during the twentieth century, the increase in cadmium concentrations in seal hair indicates that it is significantly polluting the Arctic. Among the native people who rely on hunting for their livelihood, mercury is found in very high concentrations. Such pollutants may be transported to marine environments by river outflows, atmospheric systems, and oceanic currents. Much of this problem is thought to emanate from the former Soviet Union and

Eastern European countries, although significant contamination probably originates in China and other countries to the south.

Radionuclides have been well mapped both along the Eurasian coastline and along some transects of the Arctic Ocean. The waxing and waning of contamination into and out of the Arctic over time can be traced. In the 1960s and 1970s, much contamination entered the Arctic through fallout from Soviet nuclear testing. Additional sources of strontium contamination came from the Ob River into the Kara Sea, and from the Kara Sea, it swept along the Siberian coast toward Alaska. By the 1970s, radionuclide signals from the Sellafield plant in the United Kingdom were found as far north as the Barents Sea, and by the late 1970s and 1980s, they had worked their way into the Kara Sea, merging with the dying signals from the Ob River. In the late 1980s, the Chernobyl power plant released an immense tongue of contamination into the Baltic Sea and into the Arctic by way of the Kola Peninsula. And today, the greatest signal of cesium contamination in the northern waters lies in the Baltic Sea, an indication of the drainage of radionuclides into that sea, and from the Baltic to the Norwegian-Greenland Sea.

However, levels of radionuclide contamination remain higher in the high Arctic waters than in the waters immediately adjacent to much of the Arctic coastline. Where is this enhanced concentration originating?

Organochlorines, such as the insecticide DDT, and PCBs (polychlorinated biphenyls, chemicals once commonly present in electrical generating and transformer stations and disposal sites) have dispersed over much of the Arctic, and dangerous concentrations are presently found in polar bears, other marine mammals, and humans. How can we discover the sources of these contaminants? We know that they often flow into the Arctic through the vast stretches of Arctic rivers, but we do not know how they get into these rivers or from where. Furthermore, we do not know the extent to which these chemicals arrive by way of wind or fallout that enters water, snow, and ice.

Atmospheric pollutants are present in Arctic haze, which rivals the smog of Los Angeles. This haze is made up of soot, organochlorines, heavy metals, radionuclides, CFCs (chlorofluorocarbons), sulfates, hydrocarbons, and other greenhouse gases that work their way north to the Arctic basin. Airborne lead particles from Eastern Europe have already found their way into the blood of native Greenlanders. The haze affects human health and acidifies and poisons sensitive ecosystems. It hangs over millions of square miles of the Arctic in layers as high as 25,000 feet. The effects of this haze are compounded by the Arctic flora's slow recovery rate from toxic

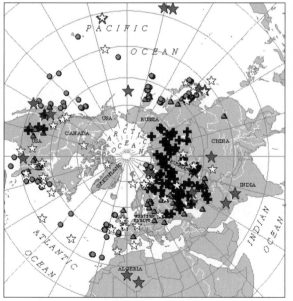

Legacy of the Cold War

fallout. For example, pine trees north of the Arctic Circle naturally take 150 to 250 years to mature, two to three times longer than pine trees in the south, even without stresses caused by pollution.

More and more, the Arctic is being looked to as the last frontier for mineral and forest extraction. Countries interested in significant Arctic resource development include Canada, Denmark, Greenland, Norway, Russia, and the United States. The aftereffects of this industrialization can be disastrous.

Environmental recovery is slower in the Arctic than in the temperate regions of the Earth. For example, oil spilled in the Arctic marine environment remains within the habitat longer than in environments with warmer climates. Decomposition is slower for several complex reasons: Low temperatures reduce evaporation and biological decomposition; long, dark winters decrease ultraviolet radiation, which can help break down the hydrocarbons; and drifting sea ice, which decreases wave action, may trap oil between and on ice floes. All these factors hinder cleanup and recovery, should an accident occur.

A threat that extends even beyond the Arctic realm is climate change. Scientists are now acutely aware of the environmental interactions between the Arctic and the overall global climate system. Most efforts to model global warming associated with worldwide emissions of carbon dioxide and other greenhouse gases reveal that temperatures will increase at least twice as much in the Arctic as in more southerly regions. Although the anthropogenic forces driving global warming lie outside the Arctic, the effects may be exacerbated by the response of the Arctic to warming. The climate of the earth depends on ocean/ice/atmosphere interactions. Even the initial stages of warming could cause the Arctic sea ice to melt, sending freshwater into the northern oceans. Abrupt freshening of the Norwegian-Greenland Sea surface water could seriously disrupt the global circulation system, which depends upon the sinking of heavy (cold, salty) water in the Norwegian-Greenland Sea. In turn, this shutting off of deepwater formation could disrupt the movement of the warm Gulf Stream, which sweeps northward from the tropics toward Europe and the Arctic, replacing the sinking surface water in that region. In an extreme scenario, Europe could be thrown into a mini–ice age.

Climate change has the potential to wreak havoc on the vulnerable Arctic marine environment. Arctic pack ice is very susceptible to minor changes in climate. Warming could set off a cascade of impacts, potentially causing dramatic changes in the way entire systems function, with significant implications for the well-being of native peoples in the region.

Recent studies show that ozone levels above the Arctic are being depleted. How might this affect the Arctic marine biosphere that lies beneath the Arctic sea ice cover? If the ice melts, organisms will be exposed to more biologically damaging ultraviolet light (UV-B rays) than normal. How severely the organisms are affected will depend not only on how quickly they can adapt but also on whether the dose they receive is within biological tolerance levels.

Large doses of ultraviolet radiation are known to kill phytoplankton, while lesser amounts slow photosynthesis. Since phytoplankton compose the base of the ocean food chain, large changes in productivity could undermine the entire ecosystem. Researchers agree that the presence of the ozone hole will result in a decline in phytoplankton production. This could lead to an even greater greenhouse effect.

The broad dispersion of these effects results from a combination of air patterns, water currents, and prevailing environmental circumstances in the Northern Hemisphere. Air currents deliver pollutants, produced largely in Eurasia, to the Arctic Basin. The low level of precipitation in the Arctic ensures that these pollutants remain in the atmosphere for lengthy periods of time, only to be dumped out to the ocean and to the ground in the spring storms, when ice is melting, rivers are flowing, and life springs back after winter. Birds and whales arrive, plankton blooms, seals feed, and fish migrate.

Clearly, a viable action plan that addresses present-day Arctic contamination is needed. The costs of such a plan most likely will be enormous. The many governments and international agencies involved will need to determine where and how to focus their various efforts.

The Arctic Environmental Protection Strategy has thus far focused on regional environmental issues and possible collaborative efforts. The Arctic Monitoring and Assessment Program (AMAP), signed in Finland in 1991, is a good foundation, but the risk is that we could fall into the trap of studying the demise of the Arctic before taking concrete steps to protect it.

Regionwide agreements are still necessary to establish procedures for carrying out environmental impact assessments and for protecting ecosystems and indigenous peoples. During the initial Finnish negotiations, many controversial issues, such as oil and gas development, were not addressed. Furthermore, the first Arctic Protection Strategy document did not contain specific proposals for follow-up actions, once data had been collected on pollution, ecosystem contamination, and global change.

In the United States, the funds for the Arctic Nuclear Waste Assessment Program have been depleted; the expeditions were funded and carried out, but there was no funding to assess the data. Only emergency funds from the Office of Naval Research allowed us to complete the *Arctic Environmental Atlas*. Despite these set-backs, the Environmental Protection Agency, the National Oceanic and Atmospheric Administration, and the Department of Energy are presently carrying out Arctic programs that are attempting to deal with chemical contaminant inventories, radionuclide waste storage, and climate change.

As early as 1991, the Environmental Defense Fund suggested that the following issues be addressed:

- the promotion of circumpolar cooperation in the protection of Arctic biodiversity
- the elimination of Arctic haze
- the negotiation of a regional sea agreement for the Arctic Ocean
- the creation of a liability regime for spills of oil and other pollutants in the Arctic
- the establishment of data exchanges between Arctic researchers and the World Climate Program
- the adoption of uniform environmental impact assessment procedures, which should be subject to an independent external review process, thereby assuring that they are based on scientifically credible data

These goals require the development of policies that focus on research, monitoring, and technology transfer to assist the Eastern European and northern Asian countries with industrial clean-up and the creation of a legal framework within which environmental degradation can be identified and corrected.

Mechanisms developed for monitoring environmental parameters should be made uniform for use among the many different nations. Equipment should be standardized and of the highest quality. Researchers from the Former Soviet Union, China, Korea, and Eastern European countries need common training together with Western scientists to operate the monitoring equipment.

A mechanism needs to be established to ease the transfer of technology among nations to provide for state-of-the-art monitoring and research equipment. The exchange of scientists and scientific information among the eight Arctic nations must function in lock-step fashion. The ultimate goal should be to retrofit or replace outdated technology in the Former Soviet Union, Eastern Europe, and Asia, thereby reducing air and water pollution at the sources.

Methods to facilitate economically advantageous and environmentally sound joint ventures among all eight Arctic nations should be encouraged. A potential joint venture is the construction of adequate nuclear waste repositories that are built, monitored, and maintained by members of all eight Arctic nations. If repositories are *not* built, the problems will only be magnified. Solutions to nuclear waste must be found not only for Russia but also for all the nuclear nations of the world, including the United States. Collaborative programs to inventory toxic chemicals also need to be broadened, and the means to destroy or contain these chemicals should be developed.

Finally, we need to address the issues of risk assessment and the ways in which scientists can impact the global environmental agenda. Gary Snyder, the 1960s leader of the environmental movement, pointed out, "It is not enough just to love nature or to want to be in harmony with nature. Our relation to the natural world must be grounded in information and experience." This statement reveals the challenge of global environmental problems. We are confronted with planetary change brought on by our own gross neglect of the atmosphere we breathe, the earth beneath our feet, and the oceans that gave life to our planet. Do we wait for the confirmation of an impending global catastrophe before we try to prevent it? We should do the opposite. This is the essence of the precautionary principle.

Barry Commoner, an American environmentalist who proposed that the Cold War and economic greed exacerbated the environmental crisis, reminded us of our intimate connectedness with the environment:

1. Everything is connected to everything else. A change in one thing can trigger an even bigger change in other things.
2. Everything must go somewhere. It does not disappear.
3. The earth's natural systems are stronger than the power of humans.
4. There is no such thing as a free lunch.
5. Everything is changing.
6. Planet earth does not recognize political boundaries.

We need to remember this.

COMPOTE

The failed *Mendeleev* expedition in 1993, which forced us to abandon our work in the Kara Sea, strained the relationships between the Naval Research Laboratory, the Office of Naval Research, and the Shirshov Institute of Oceanology. With considerable trepidation, we nevertheless initiated a new set of expeditions with the Russians, first to the Barents Sea and then to the Norwegian-Greenland Sea.

We had shifted our collaborative efforts from the Shirshov Institute to VNIIOkeangeologia in St. Petersburg upon the advice of a Norwegian colleague, Anders Solheim from the Norwegian Polar Institute. Through Anders, I had met Leonid Polyak, a St. Petersburg based scientist. Through Leonid, I learned of the Arctic capabilities of VNIIOkeangeologia. At his invitation, three U.S. representatives—David Nagel, Chris Jones, from the NRL, and I—traveled to St. Petersburg to meet with the directors of VNIIOkeangeologia.

My Russian friends will never let me forget our team's opening lines at our meeting with the director, Igor Gramberg. David Nagel started by stating, "First, let me say that we are not spies." This was probably a smart opener, since I was sure there were doubts. We quickly got into the business of discussing our shared interest in funding people to process the Barents Sea and Kara Sea contamination data. Later that night, we attended a birthday party for Leonid at his parents' apartment, even though he was absent. As on so many occasions, the vodka flowed like a river. Nagel and Leonid's father discovered that they had traveled nearly parallel paths during their careers, which, for Dave, was a personal insight. In the taxi back to the Astoria Hotel, as we careened around huge snowdrifts, Dave commented, "To think that I have spent so much of my life engaged in activities aimed at destroying these beautiful people."

The simple act of sharing a celebration, taking the time to learn about and appreciate another culture—this kind of personal, individual diplomacy—wiped out years of societal prejudice. For that alone, our efforts to further U.S. and Russian collaboration were worthwhile.

The expedition that followed was conducted on the *Geolog Fersman* under Leonid Polyak and Gennady Ivanov. One representative from the NRL, Patty Jo Burkette, participated on board, taking sediment samples. Then, in 1996, Georgy Cherkashov was appointed chief scientist for a completely new expedition on the research vessel *Professor Logachev* to map gas hydrates and the plate boundary in the Norwegian-Greenland Sea. This was the first collaborative effort by Russia, the United States, and Norway to carry out exploration in the Nordic seas. To celebrate the occasion, the U.S. ambassador to Norway visited the ship in Spitsbergen, marking a triumphal international success. In 1998, the Naval Research Lab worked again with Russia, Germany, and Norway to carry out a very difficult set of *MIR* diving operations to the bottom of the Norwegian-Greenland Sea. During this expedition, violent storms paraded in and out of our work area, turning submersible recovery into rescue operations amidst cascading waves and howling winds.

*The author, U.S. Ambassador to Norway Thomas Loftus,
Peter Vogt, U.S. Naval Attaché to Norway, and Georgy
Cherkashov discuss expedition plans on board the Russian
research vessel* Professor Logachev, *1996*
PHOTO: *AIP Nielsen*

*Kathy Crane with
Georgy Cherkashov
before a* MIR *dive,
Norwegian-
Greenland Sea,
1998*
PHOTO: *Devorah
Joseph*

In 2000, a Japanese group led by Kensaku Tamaki, from Tokyo's Ocean Research Institute joined the collaboration and assumed financial responsibility. Our menu rapidly changed from kasha and 10,000 gallons of *compote,* a Russian drink that we were sure filled the stabilizer tanks in the hull of the *Logachev,* to ramen noodles, sushi, and sake. Each member of the Japanese team wore identical black work uniforms. The team included three women scientists, a North Vietnamese scientist, and representatives from Korea, China, and Taiwan. The Chinese and Taiwanese representatives were both named Wang. They bunked together and took care of each other through storms and sickness. To me, the addition of new partners year by year illustrated how much our collaborative work had progressed in one short decade.

28

COME AWAY

Come away, O human child
To the waters and the wild
With a faery, hand in hand
For the world's more full of weeping
Than you can understand

—WILLIAM BUTLER YEATS

In the dark of the deep Siberian night, somewhere between Tomsk and Omsk, the train whistle echoed across the forested flat land. Inside the rocking cabin, the lights flickered, illuminating the frightened faces of two of the four little girls traveling to their new parents who were waiting for them in Novosibirsk. The children were bundled in thick Russian coats and hats. A nurse from the orphanage lay down on a bottom bunk trying to comfort one of the children. My sister, Ann, who braved this trip to accompany me, rested on the bunk above, and the fourth little girl, my new daughter, lay snuggled around my legs, her face like stone and her eyes as hard as the steel of the train.

It was a cold September in 1996. Winter was close. I had come to Russia to rescue this child and offer her another life. The train continued on its long journey swaying back and forth to Novosibirsk. For me, this night, which represented the long, dark transition to motherhood, had begun many years earlier, in the empty spaces of my heart that could not be filled by career, friends, or romantic involvements.

"Mama," my daughter grabbed me tightly. "Mama!"

The nurse smiled and spoke in Russian: "I could not hold her back. She came marching after you down the long corridor outside of the room of sleeping toddlers. She knew this afternoon that you were here to take her. We try to prepare all the children, to let them know that their parents have come for them. She knew that you were here for her."

That afternoon, my sister and I had managed to swallow our dinner, which had been prepared especially for us, but I could not taste a thing. I had had only a few hours to confirm, "Yes, I will take this child," and then we were on our way again, by train, the endless Trans-Siberian Railroad.

On top of my anxiety, I was suffering from the flu, experiencing double vision and extreme exhaustion. I was now relying on my sister, a mother of three, to make the right decisions for me. When we had walked into the orphanage earlier that day, all the small children were shouting, "Mama, Papa, Mama, Papa." Their cries brought tears to my eyes. And then I saw my daughter, playing with several other children. I was almost too afraid to reach out. Here was the person whose life would soon be tied to mine. I had in my hand the only likeness of my future daughter, a black and white photograph showing a little girl with a hard, somber face. But here in front of me was a child who had a quick eye, who was strong, and who smiled.

"You must agree to take her. You are so lucky!" Ann urged. Of course, I did. We had brought clothes but no shoes, and the orphanage forbid leaving with anything except what we provided. We had to persuade the orphanage to allow us to keep my daughter's boots. To this day, I still hold them as a physical memory of one of the most important moments of my life.

After dinner, I dressed my little girl in socks, warm underwear, and a dress that seemed so long for her very tiny body. She did not say a word, and I was gripped with fear. I was afraid that I would not be able to cope with a small child, and I was afraid that I was too sick to fight through all the paperwork necessary for her visa to America that was waiting for us in Moscow.

Only one year earlier, I had separated from my long-term boyfriend. We had not been able to agree on the course of our future together. I was alone again, and I realized that I wanted more in my life.

In the early 1970s, I had developed a brilliant defense mechanism that, for a woman in science, was necessary for survival. Women were expected to forgo personal lives if they wanted to succeed as oceanographers. I remember how strange I felt in 1985 when Vincent Courtillot, my colleague and Fulbright sponsor at the University of Paris, asked me, "Where are your children, Kathy?" I was really startled, and I thought for the first time, "Yes, where *are* my children?"

Just where was the American-style feminism leading us? Was it just another trap? Did it lead women to sacrifice their personal lives to become strange pseudo-men with pseudo-privileges? Why couldn't women in America have the freedom to be professionals, as well as partners and parents if they so desired?

For so many years I both fought against and bought into the social expectations that America had held for women. It was a society in which women were conspirators as well as victims. Ironically, I did not recognize the deep double-sided injustice and personal loss that I had suffered as a woman scientist in American

society until I moved to France, a country I had always considered
to be socially backward.

My sense of personal loss was perhaps compounded by the
depression that had struck after my brother's death and had muted
my resolve to succeed at all costs. Finally, by 1995, I knew that I had
to change, to give as well as receive. I began to look outward, to
engage in a truly human way, and it was during this period that I
decided to adopt a child. The long journey through seas, storms,
and emotional upheaval finally led me to my daughter.

How could I help but love my Russian colleagues? They supported me enthusiastically, arranging for Russian medical advice
about my child and translating all the pertinent Russian laws and
regulations concerning adoption. They even offered me the option
of bringing my daughter with me out to sea, so that as a single parent, I could participate on their expeditions. They personally
guided me on every step of the physical and emotional journey,
and through my child, I would be linked forever with Russia.

During the 1990s, Russia experienced incredible chaos. The
greatest social change was the enormous economic collapse, which
resulted in the destitution of millions of Russian families. This led
in turn to the abandonment of countless children, who have been
hopelessly warehoused in orphanages. I wondered if there was an
ounce of compassion in the "new" Russian government. Just
where was it leading Russia?

Russian society is as complex today as it has ever been in its history. It is a country of compassion, tainted by widespread disdain
and humiliation. I suspect that the situation was created by more
than seventy years of betrayal by the communist state. The massive slaughter of Soviet citizens by Stalin stained the country and
all of its inhabitants. The situation forced people to develop a

schizophrenic type of thinking that allowed individuals and government alike to deny their history of atrocity. Unlike Germany, Russia continues to rewrite its own history without reconciling itself with its brutal past. Living behind a veil of denial, willfully ignoring the truth of history, fosters the betrayal of the self, of the family, and of the nation as a whole.

I still wonder how many more Russians will endure the depravation of a collapsing and increasingly corrupt society before the nation reforms and begins to support its discarded, unwanted people. How can a nation of such beauty coexist with the uncaring mentality inherited from the Soviet state?

My daughter might have become a throw-away child, discarded by a chaotic Russia. I would protect and defend her in America.

NEW WORLD

Komsomolskaya Pravda, December 20, 2000

FSB CHIEF SUMS UP OPERATIONS IN 2000

An interview with Nikolai Patrushev, head of the Federal
Security Service (FSB)

In October 1999, Sutyagin, an employee of the U.S. and
Canada Institute at the Russian Academy of Sciences, was
arrested. In the course of the investigation, we revealed
the spying activities of his contact, U.S. resident Joshua
Handler, an expert in nuclear security, who is currently in
the United States.

"If it can happen to me, it can happen to anyone," Joshua Handler
told me. It was a sobering thought. In the late 1990s, the wheels
of politics had started to turn again in Russia—this time toward
renewed paranoia of foreigners and environmentalists.

What was happening now in post-Soviet Russia? How would this
affect the United States and the world? I was not unaware of the irony
of these events. The end of the Cold War had given me back my life;
I found it in the form of a beautiful little girl. Russia had embraced
me during my darkest hour, when American society would not. Russia had restored my capacity to be human. Nonetheless, I had also
learned through these experiences to be wary. In some cases, Russians had led me astray. I had confused professional relationships

with personal sincerity. More recently, the Russian government was blaming external elements for its economic demise and social dysfunction. During this period, my colleagues working in Russia and I tried to hold together our joint programs. I was still involved in the mapping of the Arctic environment. We were still able to organize expeditions on Russian ships to work on problems ranging from the investigation of gas hydrates to the study of the Arctic mid-ocean ridge system. By the time I had secured my sabbatical, I was prepared to move to Russia and to take my daughter back to her birthplace. I had arranged to share an apartment rented by my German friend, Heidi Kassens, I had arranged for office space at VNIIOkeangeologia in St. Petersburg, as well as an office at the Institute for the United States and Canada in Moscow, and most important, I had arranged for financing from the World Wildlife Fund to produce a Russian version of the *Arctic Environmental Atlas*.

Then, suddenly, my plans were turned upside down when the Russian government began to blame the problems of the country on foreign spies, many of them environmentalists. They arrested environmental activists and those with connections to the former military. Soon, the government published reports that Russian environmental organizations were filled with foreign spies.

I grew alarmed, especially for my close friends. For years, I had known and worked with Josh Handler, who had been associated with Greenpeace and then with Princeton University. In 1999, he had been living in Moscow and had an office at the Institute of the United States and Canada, the same organization where I was supposed to produce the Russian environmental atlas. Josh and I had communicated often, because he was advising me on the logistics of moving my family to Russia. I was shocked when I read in the newspaper that he had been detained by the FSB. My worst nightmare was getting arrested as a suspected environmental spy by the increasingly paranoid Russian government. Even though I was not a direct employee of the Naval Research Laboratory, would the Russians assume that my work for the NRL was somehow linked to covert operations for the U.S. government? Scientists in Russia and the Ukraine, those who were funded by the Office of Naval Research, and in some cases by NATO, were followed, questioned, and harassed. Could a situation arise that would lead to my arrest or to the arrest of those working with me?

Soon, the U.S. government responded with its own version of spy game hysteria. The FBI began to track my movements and to monitor my friends. It seemed that I would have to be wary of both governments.

After Josh was detained and questioned by the FSB, I felt that the same thing could happen to me. I decided to delay my visit to Russia. I delayed it three times more during the fall and winter of 1999 and 2000. Finally, Joan Gardner, a trusted colleague, offered to take care of my daughter, so that I could return to Russia and complete the atlas.

I made a series of visits, none of which were longer than two weeks. Over the series of visits, my Russian partners and I completed what turned out to be a beautiful atlas and delivered it to the World Wildlife Fund's Moscow headquarters, where it remains today, perhaps because the subject matter has been deemed too controversial for wide release.

Nevertheless, the publication process required a great deal of caution. We were careful to compile only data that had been published before the assembly of the atlas. This decision ensured that no government could claim that any data contained therein were bought or stolen.

Immediately after each visit to Russia, the FBI called. They also visited the NRL to question me about my Russian colleagues. I explained to them that I had no security clearance, and that nearly all of us in the Marine Geosciences division of the NRL worked with Russians.

I knew that the FBI was following me even after their visit to the NRL, and I wondered why they had suspicions of me. This uneasy situation continued for one year, until Robert Hansen, an FBI agent, was caught in action spying for Russia. Shortly thereafter, all the annoying phone calls from the FBI stopped, as Hansen went to trial. I was glad that this episode was over, yet it was a signal of the rising tide of fear between America and Russia.

This fear has led to a return to practices that were common during the Cold War. In Russia, the Communists are hoping to restore their party to power. Likewise, in the United States there are those who are working to revive Cold War competition and animosity. With this turn of events, I have renewed my commitment toward scientific and humanitarian collaboration between our two countries. For almost fifty years following World War II, the world was held in the grip of anxiety, divided by the threat of nuclear annihilation. It is our responsibility as scientists and as humans to make sure that this does not happen again.

30

DARK DÉJÀ VU

In 1999, an American woman scientist was allowed to sail on board a U.S. nuclear submarine for the first time. This occurred twenty-three years after three women had explored the Galápagos Spreading Center in the deep-diving research submersible *Alvin*. The Cold War–styled nuclear U.S. Navy was a much more impenetrable barrier for women than even industry or academia. Margo Edwards was that first woman. An outstanding leader, Edwards was the director of the Hawaiian Mapping and Research Group, and her task was to lead a scientific mission under the Arctic ice to map the previously uncharted seafloor. There wasn't a more competent person to break this gender barrier.

My connection with the U.S. nuclear submarine science program had begun ten years earlier, when Marcus Langseth, of Lamont-Doherty Earth Observatory, asked me to join the NSF-funded science team that would map the Arctic mid-ocean ridge system. He had added, much as an afterthought, that because I was a woman, I could not sail on the submarine with the rest of the male scientists. And not believing what I was hearing (it was like some dark déjà vu of my earlier life at Scripps), I swore that I would have nothing to do with the U.S. Arctic submarine program, and I continued to work with countries that *did* have room for me on their Arctic vessels, including Russia.

Eight years later, my colleague and friend Bruce Appelgate, of the University of Hawaii, convinced me that I should speak out publicly against this discrimination. He was persuasive, reminding me that I had had extensive oceanographic experience in the Arctic and that I had earned a measure of respect within the field. I was almost fifty years old, a mother, and aware, finally, that I needed to take an activist stand on this issue.

Never before had I spoken out about the discrimination toward women at sea. I merely endured the insults, hoping to change practices through my work. The speech I planned to give in 1998 would be the most difficult public talk of my life. It would be addressed to a roomful of mostly male scientists, various NSF division heads, and several admirals from the U.S. Navy, including Admiral Gaffney, who had funded my work on the Arctic atlas. It had been almost thirty years since the Equal Rights Amendment was passed by the U.S. Congress. I had worked in the Arctic for more than twenty years, but never on an American ship. Over those years, through funding from the Office of Naval Research and the National Science Foundation, I had collaborated with Swedish, Norwegian, Canadian, Russian, French, and German scientists. All of these countries provided facilities for women on their Arctic fleets. Yet, because I was a woman, I was not allowed to participate in the most important of *American* Arctic voyages, the Scientific Ice Expeditions (SCICEX) conducted aboard U.S. Navy submarines.

The room was packed when I stood up to make a long comment. My hands were shaking, and because I was so nervous, I also asked permission to read a prepared statement to the audience. Here is an excerpt:

Try to picture, in your mind, oceanography thirty years ago.

Money was easy. Rapid development of engineering and scientific pro-grams came fast and furious, aimed at the investigation of the mid-ocean ridge and all its phenomena.

Systems were designed to go on board navy vessels and nonmilitary research vessels, which consisted of a fleet of AGORs (AGOR is the name of a U.S. Navy class of ships) *donated to the oceanographic community by the Office of Naval Research.*

There were some problems, though, and these were big problems for those of us hoping and sweating to make for ourselves careers as seagoing oceanographers.

Some of us in the Scripps Deep-Tow Group were denied access to these ships, which were a fundamental platform for our missions, not because we were less intelligent, less physically capable, or less emotionally stable, but simply because we were women.

In the late 1960s, the only way that Tanya Atwater could sail on a ship to collect her thesis data was to set up a mattress underneath the plotting table of the ship's main lab. By 1973, there were four women working with the Deep-Tow Group: Karen Wishner, Marcia McNutt, Kathy Poole, and me. We routinely were denied participation on Scripps's missions that took place on board navy ships.

In the early 1970s, gender discrimination was against the law in the United States. However, the law was not often enforced, especially in the oceanographic community. To combat the situation meant risking an entire career in the process.

By the early 1970s, women at Scripps finally gained the right to sail on board the Washington *and the* Melville, *but only at a quasi-secretarial level, to keep the watch log. No women were allowed off the main deck or onto the fantail to look after, fix, or launch and retrieve their own equipment.*

Before expeditions, women were gathered in a private setting and given instructions on how they should act and what they should wear, as in heavy bulky sweaters when we sailed in the tropics. It was made very clear that a single failure would finish our careers.

Often, situations would arise where a woman could not find another woman with whom to sail. Because only pairs of women were allowed due to bunking arrangements, women routinely lost their slots on expeditions. Instead we would hear, "It's okay. We will gather your data for you and you can process it at home. It's okay, just give us your instruments, and we'll lower them where you want."

And so on. Of course much of the time, the data were not collected. On most expeditions the game plan changed while at sea, and if the woman was not there to react to the changing plans, the final results were not successful.

Many of you in this room cannot imagine how professionally and personally debilitating these situations were for the women who wanted only to be seagoing oceanographers.

Can you imagine trying to write a thesis, on top of trying to battle for your right just to sail on the ship that was needed to collect those data?

Or think of it this way. If time is U.S. taxpayers' money, then the time it took for women to achieve the right just to sail on board diminished the amount of money she had to address the scientific questions at hand. It was not only women who lost out, but taxpayers and the advance of science, too.

I was relieved that this era ended by the latter half of the 1970s.

In 1977, women gained the right to dive in the Alvin to the Galápagos Spreading Center. I was one of those women. In 1977, NASA hired the first female U.S. astronauts to fly aboard the space shuttle.

In 1980, the Office of Naval Research approached me at a conference and asked me, "Why don't you bring your sonars and camera systems up to the Arctic? It is untouched territory."

Since that time, colleagues from the Naval Research Lab, Lamont-Doherty, Scripps Institution of Oceanography, many institutions in Norway, Sweden, Germany, France, Italy, and Russia—and I—have done just that.

You can imagine my interest in the development of the SCICEX program in the early 1990s. Finally, here was a program that would allow us to crack the ice barrier, to take sonars and submarines—this time for research—into and around the Arctic.

Perhaps you can imagine also my dismay, when at lunch one day, one of my colleagues confided to me about a problem, namely, that he wanted me on the SCICEX mission, but women were not allowed to sail.

Hearing these words in the 1990s was like a dark déjà vu, or the return of the Black Plague after twenty years. But although I was offended deeply, I was also disappointed and tired, and hoped that some other braver soul would take up this cause against discrimination. Instead of fighting the situation, I chose to ignore the entire program.

Only this year, a male colleague convinced me that I should talk openly about my perspective on the SCICEX program.

It is not right that an NSF-funded program prohibits the full participation of the American oceanographic community. It is not right to ask women again to take a back seat and to be satisfied with data that "will be collected" for us.

It is also, I believe, against the law.

If I am any example, there are other professional seagoing scientists who have avoided this program because of the reasons I just mentioned and who could make it a much better program because of their expertise in their respective fields. We cannot advance as a whole when even one of us is excluded.

How can we turn this lemon into lemonade?

Maybe by doing what the navy did with the research oceanographic fleet many years ago, when it donated its AGORs to oceanographic institutions.

Instead of piggybacking a research operation onto a military vessel that can be called at any time into a "theater of war," I would suggest that the navy donate a submarine to the oceanographic community. If NSF can justify maintaining the bases, aircraft, and facilities that are the backbone of the U.S. Antarctic program, it can put its support behind an Arctic submarine program.

I can imagine the stream of objections. But if our government wants to have an Arctic Research Program, it should do it the right way and not the wrong way. A submarine "in our research pocket" would be a fantastic advance for U.S. Arctic science, but it is no advance if any members of our research community are banned from its full use. I hope that everyone in this room thinks about and acts on what I have said today. Let us change that lemon to lemonade.

At the end of my speech, there was at first a deep quiet and then a loud ovation. I was deeply moved. Admiral Gaffney walked to the back of the room and shook my hand. And he was the most courageous person present, because he took it upon himself to request a change in U.S. Navy policy. A woman scientist would sail on the next submarine expedition to the Arctic.

And I? I had a child to go home to.

Twenty-nine years have passed since I first began my explorations in oceanography. As I reflect on these decades, I realize that my struggle to surmount the difficulties that I and other women faced in science has led me to hope that our collective triumphs, as small as they may be, will lead to a more humane and equitable society.

FURTHER READING

~

Anderson, Roger. 1996. *Marine Geology.* New York: Wiley.

Associated Press. 1992. "Gates Warns of Contamination in Former Soviet Union." *Washington Post,* August 17, A7.

Ballard, Robert D., and J. Frederick Grassle. 1979. "Return to the Oases of the Deep." *National Geographic Magazine* 156, no. 5.

Ballard, Robert D., and Malcom McConnell. 2001. *Adventures in Ocean Exploration: From the Discovery of the Titanic to the Search for Noah's Flood.* Washington, D.C.: National Geographic Society.

Ballard, Robert D., and Rich Archbold. 1989. *The Discovery of the Titanic.* New York: Warner Books.

Bernton, Hal. 1993. "Russian Revelations Indicate Arctic Region Is Awash in Contaminants." *Washington Post,* May 17, A3.

Bonatti, Enrico, and K. Crane. 1983. "Anomalously Old Uplifted Crust Near Oceanic Transforms: Result of Oscillatory Spreading." *Nature* 300: 343.

Bonatti, Enrico, and K. Crane. 1984. "The Geology of Oceanic Transform Faults." *Scientific American* 250: 40.

Broad, William J. 1997. *The Universe Below: Discovering the Secrets of the Deep Sea.* New York: Simon & Schuster.

Carson, Rachel. [1951] 2003. *The Sea Around Us.* New York: Oxford University Press.

Cone, J. 1991. *Fire Under the Sea.* New York: William Morrow.

Corliss, John B., and Robert D. Ballard. 1977, "Oasis of Life in the Cold Abyss." *National Geographic Magazine* 152, no. 4.

Corliss, J., J. Dymond, L. Gordon, J. Edmond, R. P. von Herzen, R. D. Ballard, K. Green, D. Williams, A. Bainbridge, K. Crane, and T. H. van Andel. 1979. "Submarine Thermal Springs on the Galápagos Rift." *Science* 203: 1073–1083.

Courtillot, V., and G. E. Vink. 1983. "How Continents Break Up." *Scientific American* (July): 42.

Cousteau, Jacques Ives. 1985. *Jacques Cousteau and the Ocean World.* New York: Abradale Press, Harry N. Abrams.

Cox, Allan. 1973. *Plate Tectonics and Geomagnetic Reversals.* New York: W. H. Freeman.

Crane, Kathleen. 1978. "Structure and Tectonics of the Galápagos Inner Rift." *Journal of Geology* 86: 715–730.

Crane, Kathleen. 1985. "The Spacing of Rift Axis Highs: Dependence upon Diapiric Processes in the Underlying Aesthesnosphere?" *Earth, Planetary and Space Science Letters* 72: 405–414.

Crane, Kathleen. 1987. "Seafloor Mineral and Geothermal Resource: A Videotape Production to Educate High School Students on Rational Utilization of Deep-Sea Resources (Project HEAT)." *Journal of Washington Academy of Sciences* 77 (no. 4): 251.

Crane, Kathleen, and J. L. Galasso. 1999. *Arctic Environmental Atlas.* Washington, D.C., and New York: Office of Naval Research, Naval Research Laboratory, and Hunter College.

Crane, Kathleen, and R. D. Ballard. 1980. "Structure and Morphology of Geothermal Fields at the Galápagos Spreading Center 86 degrees W." *Journal of Geophysical Research* 85 (no. B3): 1443–1454.

Crane, Kathleen, and W. R. Normark. 1977. "Hydrothermal Activity and Structure of the East Pacific Rise 21 degrees N." *Journal of Geophysical Research* 82: 5336–5348.

Crane, K., J. Galasso, C. Brown, G. Cherkashov, G. Ivanov, V. Petrova, and B. Vanstayan. 2001. "Northern Ocean Inventories of Organochlorine and Heavy Metal Contamination." *Marine Pollution Bulletin* 43 (nos. 1–6): 28–60.

Davis, Devra. 2002. *When Smoke Ran Like Water: Tales of Environmental Deception and the Battle Against Pollution*. New York: Basic Books.

Earle, Sylvia A., and Joelle Delbourgo, ed. 1995. *Sea Change: A Message of the Oceans*. New York: G. P. Putnam Sons.

Fagan, Brian. 2000. *The Little Ice Age: How Climate Made History, 1300–1850*. New York: Basic Books.

Feshbach, Murray, and Alfred Friendly Jr. 1991. *Ecocide in the USSR: Health and Nature Under Siege*. New York: Basic Books.

Fleming, Fergus. 2002. *Ninety Degrees North: The Quest for the North Pole*. New York: Grove.

Ford, Corey. 1966. *Where the Sea Breaks Its Back: The Epic Story of Early Naturalist Georg Stellar and the Russian Exploration of Alaska*. Illustrated by Lois Darling. New York: Little, Brown.

Gage, John D., and Paul A. Tyler. 1993. *Deep Sea Biology: A Natural History of Organisms at the Deep Sea Floor*. New York: Cambridge University Press.

Gross, Grant. 1995. *Oceanography: A View of the Earth*, 7th ed. Upper Saddle River, N.J.: Prentice Hall.

Handler, Joshua. 1991. "Soviets Subs—A Neglected Nuclear Time Bomb." *Christian Science Monitor*, December 18.

Helvarg, David. 2001. *Blue Frontier: Saving America's Living Seas*. New York: Henry Holt.

Hollister, C. D., A.R.M. Nowell, and P. A. Jumars. 1984. "The Dynamic Abyss." *Scientific American* (March): 42.

Hoversten, Paul. 1992. "Panel to Probe Nuclear Pollution of Arctic Ocean." *USA Today*, August, 14, A3.

Hsu, Kenneth J. 1992. *Challenger at Sea*. Princeton: Princeton University Press.

Johnson, Donald S. 1995. *Charting the Sea of Darkness: The Four Voyages of Henry Hudson*. Edited by Philip Turner. New York: Kodansha International.

Kellogg, Sanoma Lee, and Elizabeth J. Kirk, eds. 1996. *Reducing Wastes from Decommissioned Nuclear Submarines in the Russian Northwest: Political, Technical, and Economic Aspects of International Cooperation*. Proceedings from the NATO

Advanced Research Workshop "Recycling, Remediation, and Restoration Strategies for Contaminated Civilian and Military Sites in the Arctic Far North," American Association for the Advancement of Science.

Kundera, Milan. 1984. *The Unbearable Lightness of Being.* Translated by Michael H. Heim. New York: Harper Collins.

Kunzig, Robert. 1999. *The Restless Sea: Exploring the World Beneath the Waves.* New York: W. W. Norton.

"Lake Baykal's Deep Vent: A Freshwater First." 1990. *National Geographic Magazine,* Geographica, December.

Lonsdale, Peter. 1977. "Abyssal Pahoehoe with Lava Coils at Galápagos Rift." *Geology* 5 (no. 3): 6.

Lonsdale, Peter. 1977. "Clustering of Suspension-Feeding Macrobenthos Near Abyssal Hydrothermal Vents at Oceanic Spreading Centers." *Deep-Sea Research* 24 (no. 9): 0.

Lonsdale, Peter. 1977. "Deep-Tow Observations at Mounds Abyssal Hydrothermal Field, Galápagos Rift." *Earth and Planetary Science Letters* 36 (no. 1): 19.

Lonsdale, Peter, and F. N. Spiess. 1977. "Abyssal Bedforms Explored with a Deeply Towed Instrument Package." *Marine Geology* 23 (nos. 1–2): 19.

Lourie, Richard. 2002. *Sakharov: A Biography.* Somerville, Mass.: Brandeis University Press.

Lutz, Richard A. 2000. "New Eyes on the Ocean's Deep Sea Vents: Science at the Extreme." *National Geographic Magazine* 198, no. 4.

Macdonald, K. C., and B. P. Luyendyk. 1981. "The Crest of the East Pacific Rise." *Scientific American* (May): 100.

Marshall, Elliot. 1992. "A Scramble for Data on Arctic Radioactive Dumping." *Science* 257, July 31.

Maugh II, Thomas H. 1990. "Bottom of Siberian Lake a Biological Treasure Trove." *Los Angeles Times,* August 10, A1.

McKenzie, D. P. 1983. "The Earth's Mantle." *Scientific American* (September): 66.

McPhee, John. 1993. *Assembling California.* New York: Farrar, Straus & Giroux.

Roland Huntford. 1999. Introduction to *Farthest North,* by Fridjof Nansen. New York: Modern Library.

O'Brien, Flann. 1967. *The Third Policeman.* New York: Walker and Company, Penguin Group.

O'Brien, Flann. 1997. *The Dalkey Archive.* Normal, Ill.: Dalkey Archive Press.

O'Brien, Flann. 1998. *At Swim–Two–Birds.* Normal, Ill.: Dalkey Archive Press.

Oreskes, Naomi, ed. 2001. *Plate Tectonics: An Insider's History of the Modern Theory of the Earth.* Boulder: Westview Press.

Parker, Tim. 1992. "Russian Nuclear Mess Threatens Arctic." *Fairbanks Daily News-Miner,* August 16.

Peary, Robert E., T. Roosevelt, and Robert M. Bryce. 2001. *The North Pole: Its Discovery in 1909 Under the Auspices of the Peary Arctic Club.* New York: Cooper Square Press.

Pfirman, Stephanie, Kathleen Crane, and Peter deFur. 1994. "Arctic Contaminant Distribution, CARC." *Northern Perspectives* 21, no. 4.

Raitt, Helen. 1956. *Exploring the Deep Pacific.* New York: W. W. Norton.

Rhodes, Richard. 1987. *The Making of the Atomic Bomb.* New York: Simon & Schuster.

Safina, Karl. 1998. *Song for a Blue Ocean: Encounters Along the World's Coasts and Beneath the Seas.* New York: Henry Holt.

Schiller, Ulrich. 1993. "Versenkte Zeitbomben." *Die Zeit* 27, July 2.

Sclater, J. G., and C. Tapscott. 1979. "The History of the Atlantic." *Scientific American* (June): 156–174.

Seibold, E., and W. H. Berger. 1996. *The Sea Floor,* 3rd ed. New York: Springer-Verlag.

Shakespeare, William. 1998. *The Tempest.* Reprint, edited by Stephen Orgel. Oxford World's Classics. New York: Oxford University Press.

"Soviets Reported to Have Dumped Nuclear Waste in Arctic Waters." 1992. *Boston Globe,* February 26.

Steller, Georg Wilhelm. 1988. *The Journal of a Voyage with Bering, 1741–1742.* Palo Alto: Stanford University Press.

Sullivan, Walter. 1977. "Hot Springs on Ocean Floor Found Teeming with Life." *New York Times,* April 19.

Thoreau, Henry David. 1995. *Walden.* Boston: Houghton-Mifflin.

Van Dover, Cindy Lee. 1996. *Deep-Ocean Journeys: Discovering New Life at the Bottom of the Sea.* Cambridge, Mass.: Perseus Books.

Van Dover, Cindy Lee. 2000. *The Ecology of Deep-Sea Hydrothermal Vents.* Princeton: Princeton University Press.

Verne, Jules. 1988. *Twenty Thousand Leagues Under the Sea.* New York: Oxford University Press.

Weiss, Ray, P. Lonsdale, J. E. Lupton, A. E. Bainbridge, and H. Craig. 1977. "Hydrothermal Plumes in the Galápagos Rift." *Nature* 267 (no. 5612): 4.

Wilford, John Noble. 1990. "Soviet Lake Offers Look at Evolution." *New York Times,* August 10, A6.

Wolfe, Tom James. 1983. *The Right Stuff.* New York: Farrar, Straus & Giroux.

Zachary, G. Pascal. 1997. *Endless Frontier: Vannevar Bush, Engineer of the American Century.* New York: Free Press.

INDEX